369 0246882

KT-388-142

Adipose Tissue and Cancer

Mikhail G. Kolonin
Editor

Adipose Tissue and Cancer

Editor
Mikhail G. Kolonin
Institute of Molecular Medicine
Center for Stem Cell and Regenerative Medicine
University of Texas Health Science Center at Houston
Houston, TX, USA

ISBN 978-1-4614-7659-7 ISBN 978-1-4614-7660-3 (eBook)
DOI 10.1007/978-1-4614-7660-3
Springer New York Heidelberg Dordrecht London

Library of Congress Control Number: 2013943692

Printed on acid-free paper

Springer is part of Springer Science+Business Media (www.springer.com)

Preface

Obesity, the medical condition caused by white adipose (fat) tissue overgrowth, is a clear risk factor for cardiovascular disease and diabetes. Over the past few years, epidemiological studies have revealed the association between obesity and increased risk of certain cancers, as well as poor prognosis of a number of cancers.

The pathophysiology underlying the relationship between obesity and cancer is complex and incompletely understood. Until now, it has been unclear if excess fat tissue itself affects cancer progression or if this link is predominantly due to diet and lifestyle of obese individuals. Recent studies show that the state of obesity can accelerate tumor growth irrespective of diet. Based on the apparent link between increased adiposity and several cancers (colorectal, endometrial, breast, and prostate), it has been proposed that adipose tissue has a direct effect on tumors. An emerging body of evidence confirms that this cross talk indeed takes place at several levels.

Adipose tissue is composed of adipocytes, as well as vascular and stromal cells, secreting numerous soluble factors collectively termed adipokines. In addition, infiltration of the immune system cells in obesity leads to increased production of a number of inflammatory cytokines by adipose tissue, thus contributing to the establishment of the metabolic syndrome. Endocrine signaling by adipose tissue-derived molecules has been shown to promote cancer in animal models, matching clinical associations. Recent studies have shown that cells from adipose tissue are capable of trafficking to tumors, thus enabling paracrine action of adipokines from within the tumor microenvironment. Investigation of the molecular pathways through which adipose cells traffic to tumors and execute their functions is underway. Extracellular matrix modulation, increased tumor vascularization, immune system suppression, and direct effects on malignant cell survival and proliferation have been investigated as potential activities of systemic and locally produced adipokines within the tumor.

The book comes at a very timely moment as it discusses the clinical and experimental data pointing to the role of individual components of obesity in cancer and evaluates individual mechanisms through which adipose tissue excess or restriction could influence cancer progression. An introductory chapter overviews metabolic changes taking place in obesity. Next several chapters discuss the clinical data related to specific cancers promoted by obesity. This is followed by the chapters focusing on molecular players linking obesity and tumor physiology, as well as animal models to study them. Recent discoveries of the roles that host cells from adipose tissue play in tumor microenvironment are covered in the next two chapters. Finally, perspectives for obesity management as an approach to cancer prevention and treatment are discussed.

Houston, TX, USA Mikhail G. Kolonin

Contents

Chapter 1
Metabolic Perturbations Associated with Adipose Tissue Dysfunction and the Obesity–Cancer Link

Nikki A. Ford, John DiGiovanni, and Stephen D. Hursting

Abstract Nearly 35 % of adults and 20 % of children in the USA are obese, defined as a body mass index (BMI) $\geq$30 kg/m^2. Obesity, which is accompanied by metabolic dysregulation often manifesting in the metabolic syndrome, is an established risk factor for many cancers. Within the growth-promoting, proinflammatory environment of the obese state, crosstalk between adipocytes, macrophages, and epithelial cells occurs via obesity-associated hormones, cytokines, and other mediators that may enhance cancer risk and/or progression. This chapter synthesizes the evidence on key biological mechanisms underlying the obesity–cancer link, with particular emphasis on the relative contributions of increased adiposity per se versus the obesity-associated enhancements in growth factor signaling, inflammation, and vascular integrity processes resulting from adipose tissue dysfunction. These interrelated pathways represent possible mechanistic targets for disrupting the obesity–cancer link.

N.A. Ford
Department of Nutritional Sciences, University of Texas, Austin, TX, USA

J. DiGiovanni
College of Pharmacy, University of Texas, Austin, TX, USA

Department of Nutritional Sciences, University of Texas, Austin, TX, USA

S.D. Hursting (✉)
Department of Nutritional Sciences, University of Texas, Austin, TX, USA

Department of Molecular Carcinogenesis, University of Texas-MD Anderson Cancer Center, Smithville, TX, USA
e-mail: shursting@mail.utexas.edu

M.G. Kolonin (ed.), *Adipose Tissue and Cancer*, DOI 10.1007/978-1-4614-7660-3_1,

1.1 Introduction

The prevalence of obesity, defined as a body mass index (BMI) $\geq$30 kg/m^2, has increased dramatically in recent decades in the USA, and nearly 35 % of adults and 20 % of children are now obese [1]. Worldwide, an estimated 1.1 billion adults are overweight and 500 million adults are obese (http://www.iaso.org/resources/aboutobesity/). The obese state is characterized by an excessive expansion of adipose tissue mass, which manifests as adipocyte hypertrophy (increased size), hyperplasia (increased number), and increased intracellular lipids. Excessive adiposity per se can exert untoward structural and biomechanical effects on organs (such as the lungs, liver, and pancreas), blood vessels, musculoskeletal system, and other tissues [2]. In addition, the resulting adipocyte hyperplasia and hypertrophy are associated with adipocyte dysfunction that can trigger local and systemic changes characteristic of the metabolic syndrome that increase risk and worsen prognosis of several cancers and other chronic diseases [3] (Fig. 1.1).

Among obese adults, ~60 % meet the criteria for the metabolic syndrome, a state of metabolic dysregulation characterized by insulin resistance, hyperglycemia, dyslipidemias (particularly hypertriglyceridemia), and hypertension [4]. In obesity and/or metabolic syndrome, alterations also occur in circulating levels of insulin, bioavailable insulin-like growth factor (IGF)-1, adipokines (e.g., leptin, adiponectin, and monocyte chemotactic factor), inflammatory factors (e.g., cytokines), and vascular integrity-related factors [e.g., vascular endothelial growth factor (VEGF) and plasminogen activator inhibitor (PAI)-1] [5, 6]. Through these mediators, obesity and metabolic syndrome are linked to various chronic diseases [6, 7] including cardiovascular disease, type II diabetes, and the focus of this chapter, cancer. Importantly, not all obese individuals develop the metabolic dysregulation usually associated with obesity and metabolic syndrome, and these "metabolically healthy obese" individuals do not have elevated cancer risk. An estimated 30 % of obese

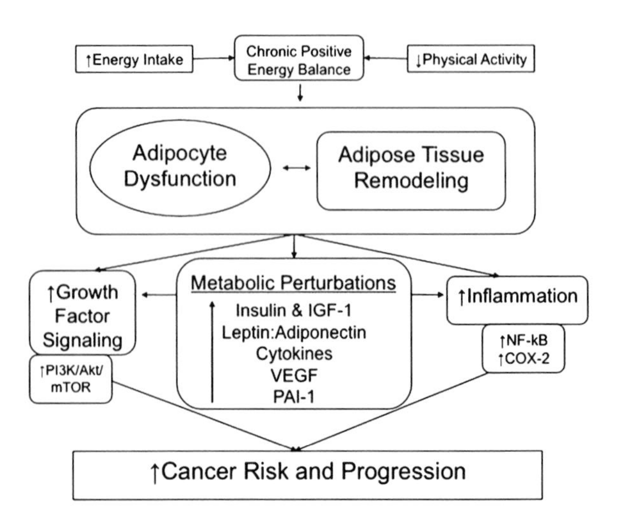

Fig. 1.1 Obesity, adipocyte dysfunction, and cancer: overview of mechanisms. An *arrow* preceding text denotes a directional effect (e.g., activity or concentration). *IGF-1* insulin-like growth factor-1, *VEGF* vascular endothelial growth factor, *PAI-1* plasminogen activator inhibitor-1, *PI3K* phosphoinositide 3-kinase, *mTOR* mammalian target of rapamycin, *NF-kB* nuclear factor kB, *COX-2* cyclooxygenase-2

individuals in the USA are metabolically healthy [8]. Conversely, some nonobese individuals can develop the metabolic perturbations usually associated with obesity, and these individuals appear to be more prone to chronic diseases including cancer [9]. Thus, an emerging hypothesis is that the obesity-related metabolic perturbations, and not specific dietary components or increased adiposity, are at the crux of the obesity–cancer connection. A central goal of this chapter is to identify the primary targets for breaking the obesity–cancer link and to weigh the evidence regarding the importance of obesity-associated adiposity versus the metabolic changes that typically accompany the obese state.

Evidence-based guidelines for cancer prevention urge maintenance of a lean phenotype [10]. Overall, an estimated 15–20 % of all cancer deaths in the USA are attributable to overweight and obese body types [11]. Obesity is associated with increased mortality from cancer of the prostate and stomach in men; breast (postmenopausal), endometrium, cervix, uterus, and ovaries in women; and kidney (renal cell), colon, esophagus (adenocarcinoma), pancreas, gallbladder, and liver in both genders [11]. While the relationships between metabolic syndrome and specific cancers are less well established, first reports from the Metabolic Syndrome and Cancer Project, a European cohort study of ~580,000 adults, confirm associations between obesity (or BMI) in metabolic syndrome and risks of colorectal, thyroid, and cervical cancer [12]. With the increasing prevalence of obesity and metabolic syndrome, strategies to break the links between these conditions and cancer are urgently needed [6].

Herein, we discuss possible mechanisms underlying the links between obesity, metabolic syndrome, and cancer, with emphasis on obesity-associated enhancements in adipocyte dysfunction, inflammation, growth factor signaling, and vascular integrity. Emphasis will be placed on the crosstalk between adipocytes, macrophages, endothelial cells, and epithelial cells in many cancers. Specifically, we describe the dysregulation of growth signals (including insulin, IGF-1, downstream signaling pathways, and adipokines), cytokines, and cellular crosstalk, and vascular integrity factors in the obese state that may contribute to multifactorial enhancement of cancer processes. Components of these interrelated pathways offer possible mechanism-based targets for the prevention and control of cancers related to, or caused by, excess body weight and the metabolic syndrome.

1.2 Obesity and Adipose Tissue Dysfunction and Remodeling

1.2.1 Adipose Tissue Function

Adipose tissue is loosely characterized as fat depots with widely varying structure, size, and function but in general serves to store neutralized triacylglycerides, which can then be released into the blood stream by lipolysis during fasting or times of heightened energy requirements [13]. Typically, in response to low energy states, cytosolic lipolysis within adipocytes releases free fatty acids into the blood stream, which can be used for beta-oxidation by peripheral tissues. In a diseased state (type 2 diabetes, metabolic syndrome, and fatty liver disease), adipose tissue fails to appropriately respond to changes in nutritional requirements resulting in altered metabolic signaling characterized by elevated adipokine and cytokine release into the plasma. During obesity, adipose tissue responds to the excess energy by increasing adipocyte size (hypertrophy) and enhancing adipocyte proliferation (hyperplasia) [14]. Adipocyte size strongly correlates with insulin resistance and secretion of proinflammatory cytokines [3]. Moreover, location of the adipose tissue also determines risk for metabolic diseases. In particular, increased visceral adipose (adipocytes located around internal organs) mass, is positively associated with insulin resistance likely because it is more metabolically active characterized by lipolytic release of more free fatty acids into the blood stream than other adipose depots (i.e., subcutaneous) [13].

1.2.2 Hypoxia and Inflammation

Healthy adipose tissue must be able to rapidly respond to excess energy intake by inducing adipocyte hypertrophy and hyperplasia, remodeling of the extracellular matrix, and enhanced neovascularization to nourish the adipose tissue. In pathological states such as insulin resistance associated with obesity, rapid adipocyte hypertrophy occurs with restricted angiogenesis resulting in cellular hypoxia, and thereby resulting in local inflammation [15]. Macrophages surrounding necrotic adipocytes phagocytize fatty acids, which are released from the adipocyte. This produces bloated, lipid overburdened macrophages, which is characteristic of chronic inflammation and often observed in obese individuals [14]. Furthermore, macrophages are predominantly differentiated to an M1 proinflammatory phenotype, which has been positively associated with systemic insulin resistance [14]. Hypoxia modulates the production of several inflammation-related cytokines and adipokines, including increased IL-6, leptin, and macrophage migratory inhibition factor production together with reduced adiponectin synthesis [15]. Increased glucose transport into adipocytes is also observed with hypoxia, largely as a result of the upregulation of GLUT-1 expression, consistent with changes in cellular glucose metabolism [15]. Crosstalk between infiltrating macrophages and adipocytes creates a positive feedback loop for further production of inflammatory cytokines within adipose tissue. Taken together, hypoxia appears to be a critical factor in the underlying cause of adipose tissue dysfunction resulting in many of the metabolic and inflammatory perturbations associated with obesity.

1.2.3 Ectopic Fat and Adipose Tissue Dysfunction

When lipid storage capacity in adipose tissue is exceeded, surplus lipids often accumulate within muscle, liver, and pancreatic tissue [16]. As a consequence, hepatic and pancreatic steatosis can develop; both have been positively associated with insulin resistance and ultimately lead to impairment of lipid processing and clearance within these tissues [16]. As a result of lipotoxic and inflammation-mediated adipocyte dysfunction, the liver and pancreas are unable to cope with the overflow of lipids and lipotoxic effects of free fatty acids [17]. Consequently, lipid intermediates impair function of cellular organelles and cause release of cytokines, which foster insulin resistance by activating phosphokinases, thus impairing insulin receptor signaling. Evidence is accumulating that pancreatic adipocyte infiltration and fat accumulation may be one of the earliest steps in obesity-associated pancreatic endocrine dysfunction, and this can trigger pancreatic steatosis (PS), nonalcoholic fatty pancreatic disease (NAFPD), and pancreatitis [5, 7]. Moreover, fatty pancreas has been positively associated with visceral adipose tissue mass and systemic insulin resistance [18]. Together pancreatic steatosis and fatty pancreas contribute to the broader metabolic and inflammatory perturbations associated with obesity and metabolic syndrome.

The term nonalcoholic fatty liver disease (NAFLD) refers to a disease spectrum that includes variable degrees of simple steatosis, nonalcoholic steatohepatitis (NASH), and cirrhosis [19, 20]. Simple steatosis is benign, whereas NASH is defined by the presence of hepatocyte injury, inflammation, and/or fibrosis, which can lead to cirrhosis, liver failure, and hepatocellular carcinoma. NAFLD is diagnosed when liver fat content is >5–10 % by weight in the absence of excess alcohol consumption or other liver disease [21]. NASH occurs in 20 % of cases of NAFLD and ~5–20 % of NASH cases progress to cirrhosis; 80 % of cryptogenic cirrhosis cases present with NASH [22]. Of this group, ~0.5 % will eventually progress to hepatocellular carcinoma, although in the presence of hepatitis C the risk increases [22]. NAFLD is one of the most common chronic diseases and in the USA and Europe [23, 24] the incidence in adults and children is rising rapidly [25, 26]. Pediatric NASH is also a global problem and the prevalence of fatty liver disease has increased concomitantly with the increase in pediatric obesity during the past 30 years. At least part of the increase may be attributed to increased recognition of this condition and NAFLD can be considered the hepatic manifestation of metabolic syndrome [27].

In Western populations, overnutrition/obesity is the most common cause of NAFLD, with an estimated incidence of 15–20 %, and an increasing number of patients presenting risk factors for its development [28–31]. Overnutrition- and obesity-related NAFLD is a multifactorial disorder and is linked to hypertriglyceridemia, obesity, and insulin resistance, as observed in patients with metabolic syndrome [32]. There is no single causal explanation for the development of primary hepatic steatosis and for how intrahepatic lipid accumulation leads to the development of inflammation. Initially, Day and James [33] proposed a "two-hit" model claiming that the reversible intracellular deposition of triacylglycerols ("first hit") leads to metabolic and molecular alterations that sensitize the liver to the second "hit," i.e., oxidative stress and subsequent activation of inflammatory pathways, cellular dysfunction, and lipoapoptosis [33]. Hepatic steatosis ultimately leads to impairment of lipid processing and clearance in the liver. Lipotoxic and inflammation-mediated mechanisms have been suggested to be responsible for adipocyte dysfunction and modulation of peripheral lipid storage capacities, which result in release of free fatty acids and hepatic lipid burdening [17]. In NAFLD, the liver is unable to cope with overflow of lipids and lipotoxic effects of free fatty acids and lipid intermediates impair function of liver cell organelles by mechanisms that involve production of ROS, ER stress, activation of proinflammatory defense programs, and eventually apoptosis. Toxic lipids and release of cytokines foster insulin resistance by activating JNK, PKCε, PKCζ, and other phosphokinases to impair IRS-1 and IRS-2 signaling. This disturbed insulin signaling contributes to diminished fatty acid oxidation as well as VLDL assembly and secretion in the liver, involving an inadequate regulation of PPARα and PPARγ [24, 34, 35]. Induction of PPARα and lipid droplet-associated proteins in the liver enable formation of lipid droplets that incorporate various lipids and provide storage for de novo synthesized triglycerides. Activation of cellular defense programs, specifically activation of NF-κB, is an important determinant for disease progression from steatosis to NASH [36].

1.3 Obesity and Metabolic Perturbations

1.3.1 Dysregulated Growth Signals

Insulin and IGF-1

Insulin is a peptide hormone produced by pancreatic beta cells and released in response to elevated blood glucose. Hyperglycemia, a hallmark of metabolic syndrome, is associated with insulin resistance, aberrant glucose metabolism, chronic inflammation, and the production of other metabolic hormones such as IGF-1, leptin, and adiponectin [37]. Sharing ~50 % sequence homology with insulin, IGF-1 is a peptide growth factor produced primarily by the liver following stimulation by growth hormone. IGF-1 regulates growth and development of many tissues, particularly prenatally [38]. IGF-1 in circulation is typically bound to IGF-binding proteins (IGFBPs) that regulate the amount of free IGF-1 bioavailable to bind to the IGF-1 receptor (IGF-1R) and elicit growth or survival signaling [38]. In metabolic syndrome, the amount of bioavailable IGF-1 increases, possibly via hyperglycemia-induced suppression of IGFBP synthesis and/or hyperinsulinemia-induced promotion of hepatic growth hormone receptor expression and IGF-1 synthesis [37]. Elevated circulating IGF-1 is an established risk factor for many cancer types [38, 39].

Signaling Pathways Downstream of the Insulin Receptor and IGF-1R

The phospatidylinositol-3 kinase (PI3K)/Akt pathway, downstream of the insulin receptor and IGF-1R, is one of the most commonly altered pathways in epithelial cancers [40]. This pathway integrates intracellular and environmental cues, such as growth factor concentrations and nutrient availability, to regulate cellular survival, proliferation, protein translation, and metabolism. Activation of receptor tyrosine kinases, such as the insulin receptor or IGF-1R, stimulates PI3K to produce lipid messengers that facilitate activation of the Akt cascade. Akt regulates the mammalian target of rapamycin (mTOR) [41], which regulates cell growth, cell proliferation, and survival through downstream mediators. mTOR activation is inhibited by increased AMP-activated kinase (AMPK) under low nutrient conditions [42]. Increased activation of mTOR is common in tumors and many normal tissues from obese and/or diabetic mice [43], and specific mTOR inhibitors block the tumor-enhancing effects of obesity in mouse models [44, 45]. Furthermore, both rapamycin (mTORC1 inhibitor) and metformin (AMPK activator) have been shown to block tumor formation in multiple animal models [46–50]. Interestingly, in some model systems, rapamycin has been shown to block inflammation associated with tumor formation [49].

Leptin, Adiponectin, and the Leptin-to-Adiponectin Ratio

Leptin is a peptide hormone produced by adipocytes, is positively correlated with adipose stores and nutritional status, and functions as an energy sensor to signal the brain to reduce appetite. In the obese state, adipose tissue overproduces leptin, and the brain no longer responds to the signal. Insulin, glucocorticoids, tumor necrosis factor-alpha (TNF-α), and estrogens all stimulate leptin release [51]. Leptin has direct effects on peripheral tissues, indirect effects on hypothalamic pathways, and modulates immune function, cytokine production, angiogenesis, carcinogenesis, and other biological processes [51]. The leptin receptor has similar homology to class I cytokine receptors that signal through the janus kinase and signal transduce rand activator of transcription (JAK/STAT) pathway that is often dysregulated in cancer [52].

Adiponectin is a hormone mainly secreted from visceral adipose tissue. Levels of adiponectin, in contrast with leptin, negatively correlate with adiposity. Adiponectin functions to counter the metabolic program associated with obesity and hyperleptinemia by modulating glucose metabolism, increasing fatty acid oxidation and insulin sensitivity, and decreasing production of inflammatory cytokines [53]. The possible mechanisms through which adiponectin exerts anticancer effects may include increasing insulin sensitivity, and decreasing insulin/IGF-1 and mTOR signaling via activation of AMPK. Adiponectin also reduces proinflammatory cytokine expression via inhibition of the nuclear factor kappa-light-chain-enhancer of activated B-cells (NF-kB) [54, 55].

Evidence from a variety of studies (including cell culture, animal, and epidemiologic) linking leptin [56–58] or adiponectin [54, 59–61] individually to cancer risk is mixed. Associations between the adiponectin-to-leptin ratio and the metabolic syndrome [62–64] and some cancers [65–67] are reported. Further characterization of these links is needed.

1.3.2 *Chronic Inflammation*

Cytokines and Crosstalk Between Adipocytes, Macrophages, and Epithelial Cells

Obesity and metabolic syndrome are associated with a low-grade, chronic state of inflammation characterized by increased circulating free fatty acids and chemoattraction of immune cells (such as macrophages that also produce inflammatory mediators) into the local milieu [68–70]. These effects are further amplified by the release of inflammatory cytokines such as interleukin (IL)-1β, IL-6, TNF-α, and monocyte chemoattractant protein (MCP)-1. Adipocytes can enlarge past the point of effective oxygen diffusion, which results in hypoxia and eventually necrosis. Free fatty acids escape the engorged/necrotic adipocytes and deposit in other tissues, which in turn promotes insulin resistance and diabetes (through downregulation of insulin receptors and glucose transporters), hypertension, and fatty liver disease and also activates signaling molecules involved in epithelial carcinogenesis such as NF-kB [71].

NF-κB is a transcription factor that is activated in response to bacterial and viral stimuli, growth factors, and inflammatory molecules (e.g., TNF-α, IL-6, and IL-1β) and is responsible for inducing gene expression associated with cell proliferation, apoptosis, inflammation, metastasis, and angiogenesis. Activation of NF-κB is a common characteristic of many tumors and is associated with insulin resistance and elevated circulating levels of leptin, insulin, and/or IGF-1 [72, 73].

Inflammation and Cancer

The link between chronic inflammation and cancer development was first noticed nearly 150 years ago by Rudolph Virchow when he observed an abundance of leukocytes in neoplastic tissue [74]. Now, inflammation is a recognized hallmark of cancer, and growing evidence continues to indicate that chronic inflammation is associated with increased cancer risk [75–77]. Several tissue-specific inflammatory lesions are established neoplastic precursors for invasive cancer, including gastritis for gastric cancer, inflammatory bowel disease for colon cancer, and pancreatitis for pancreatic cancer [78, 79].

Tumor microenvironments are composed of multiple cell types including epithelial cells, fibroblasts, mast cells, and cells of the innate and adaptive immune system [78, 80]. As discussed previously, macrophages, which are activated in the obese state, infiltrate tumors and amplify the inflammatory tumor microenvironment, often through NF-kB-dependent production of cytokines and angiogenic factors [78]. Another important cancer-related inflammatory mediator is cyclooxygenase (COX)-2, an enzyme that is upregulated in most tumors and catalyzes the synthesis of the potent inflammatory lipid metabolite, prostaglandin E_2. COX-2 overexpression is an indicator of poor prognosis in multiple cancer types [81].

In some cancers, inflammatory conditions precede malignant changes (as previously mentioned); whereas, in other cancer types, genetic alterations and premalignant changes precede the inflammatory microenvironment and neoplasia [76]. Malignancies may thus be initiated or exacerbated by inflammation, and increased levels of inflammation markers may be a cause and/or consequence of cancer [76, 77]. In either scenario, the inflammatory microenvironment exerts tumor-promoting effects, with dysregulated inflammation pathways implicated in genetic instability and also cell proliferation, survival, angiogenesis, and metastasis associated with cancer [75, 77, 82].

1.3.3 Changes in Vascular Integrity Factors

Vascular Endothelial Growth Factor

VEGF, a heparin-binding glycoprotein produced by adipocytes and tumor cells, has angiogenic, mitogenic, and vascular permeability-enhancing activities specific for endothelial cells [83]. Circulating levels of VEGF are increased in obese, relative to lean, humans and animals, and increased tumoral expression of VEGF is associated with poor prognosis in several obesity-related cancers [84]. The need for nutrients and oxygen triggers tumor cells to produce VEGF, which leads to the formation of new blood vessels to nourish the rapidly growing tumor and may facilitate the metastatic spread of tumors cells [83].

Adipocytes communicate with endothelial cells by producing a variety of proangiogenic and vascular permeability-enhancing factors. These include VEGF, IGF-1, PAI-1, leptin, hepatocyte growth factor, and fibroblast growth factor-2 [85]. In the obese, nontumor setting, these factors stimulate neovascularization in support of the expanding fat mass. These adipose-derived factors may also contribute to obesity-associated enhancement of tumor angiogenesis. Bevacizumab-based therapy (i.e., anti-VEGF therapy), in combination with conventional chemotherapy, is considered a first-line treatment option for patients with advanced colorectal cancer; however, decreased efficacy in obese patients is reported and speculated to be associated with increased levels of VEGF (and/or other proangiogenic factors) produced by visceral white adipose tissue [86, 87]. The relative contributions of tumor-derived versus adipocyte-derived, proangiogenic factors in tumor development, progression, and metastasis remain unclear.

PAI-1

PAI-1 is a serine protease inhibitor produced by endothelial cells, stromal cells, and adipocytes in visceral white adipose tissue [88]. Increased circulating PAI-1 levels, frequently found in obese subjects, are associated with increased risk of atherogenesis and cardiovascular disease, diabetes, and several cancers [5, 88]. PAI-1, through its inhibition of urokinase-type and tissue-type plasminogen activators, regulates fibrinolysis and integrity of the extracellular matrix. PAI-1 is also involved in angiogenesis and thus may contribute to obesity-driven tumor cell growth, invasion, and metastasis [5]. Although PAI-1 levels in obese individuals may be reduced via weight loss or TNF-α blockade [89, 90], the role of PAI-1 in tumorigenesis remains controversial [88].

1.4 Conclusion

Adipose tissue dysfunction, along with multiple hormones, growth factors, cytokines, and other mediators associated with the obese state, enable crosstalk between macrophages, adipocytes, endothelial cells, and epithelial cells and contribute to cancer-related processes (including growth signaling, inflammation, and vascular alterations). Components of these interrelated pathways represent promising mechanism-based targets (analogous to reducing cholesterol levels to reduce heart disease risk) for lifestyle or pharmacologic interventions to prevent or control cancer in obese or otherwise metabolically dysregulated individuals.

References

1. Flegal KM, Carroll MD, et al. Prevalence of obesity and trends in the distribution of body mass index among US adults, 1999–2010. JAMA. 2012;307(5):491–7.
2. Grassi G, Seravalle G, et al. Structural and functional alterations of subcutaneous small resistance arteries in severe human obesity. Obesity (Silver Spring). 2010;18(1):92–8.
3. Gottschling-Zeller H, Birgel M, et al. Depot-specific release of leptin from subcutaneous and omental adipocytes in suspension culture: effect of tumor necrosis factor-alpha and transforming growth factor-beta1. Eur J Endocrinol. 1999;141(4):436–42.
4. Ford ES, Li C, et al. Prevalence and correlates of metabolic syndrome based on a harmonious definition among adults in the US. J Diabetes. 2010;2(3):180–93.
5. Carter JC, Church FC. Obesity and breast cancer: the roles of peroxisome proliferator-activated receptor-gamma and plasminogen activator inhibitor-1. PPAR Res. 2009;2009:345320.
6. Hursting SD, Berger NA. Energy balance, host-related factors, and cancer progression. J Clin Oncol. 2010;28(26):4058–65.
7. Poirier P, Giles TD, et al. Obesity and cardiovascular disease: pathophysiology, evaluation, and effect of weight loss. Arterioscler Thromb Vasc Biol. 2006;26(5):968–76.
8. Bluher M. Are there still healthy obese patients? Curr Opin Endocrinol Diabetes Obes. 2012; 19(5):341–6.
9. Marques-Vidal P, Pecoud A, et al. Normal weight obesity: relationship with lipids, glycaemic status, liver enzymes and inflammation. Nutr Metab Cardiovasc Dis. 2010;20(9):669–75.
10. AICR. World Cancer Research Fund/American Institute for Cancer Research. Food, nutrition, physical activity, and the prevention of cancer: a global perspective. Washington, DC: AICR; 2007.
11. Calle EE, Rodriguez C, et al. Overweight, obesity, and mortality from cancer in a prospectively studied cohort of U.S. adults. N Engl J Med. 2003;348(17):1625–38.

12. Stocks T, Borena W, et al. Cohort profile: the metabolic syndrome and cancer project (Me-Can). Int J Epidemiol. 2010;39(3):660–7.
13. Bjørndal B, Burri L, et al. Different adipose depots: their role in the development of metabolic syndrome and mitochondrial response to hypolipidemic agents. J Obes. 2011;2011:490650.
14. Sun K, Kusminski CM, et al. Adipose tissue remodeling and obesity. J Clin Invest. 2011;121(6):2094–101.
15. Wood IS, de Heredia FP, et al. Cellular hypoxia and adipose tissue dysfunction in obesity. Proc Nutr Soc. 2009;68(4):370–7.
16. Henry SL, Bensley JG, et al. White adipocytes: more than just fat depots. Int J Biochem Cell Biol. 2012;44(3):435–40.
17. Anderson N, Borlak J. Molecular mechanisms and therapeutic targets in steatosis and steatohepatitis. Pharmacol Rev. 2008;60(3):311–57.
18. Lee JS, Kim SH, et al. Clinical implications of fatty pancreas: correlations between fatty pancreas and metabolic syndrome. World J Gastroenterol. 2009;15(15):1869–75.
19. Kotronen A, Westerbacka J, et al. Liver fat in the metabolic syndrome. J Clin Endocrinol Metab. 2007;92(9):3490–7.
20. Neuschwander-Tetri BA, Caldwell SH. Nonalcoholic steatohepatitis: summary of an AASLD Single Topic Conference. Hepatology. 2003;37(5):1202–19.
21. Vanni E, Bugianesi E, et al. From the metabolic syndrome to NAFLD or vice versa? Dig Liver Dis. 2010;42(5):320–30.
22. Adams LA, Lymp JF, et al. The natural history of nonalcoholic fatty liver disease: a population-based cohort study. Gastroenterology. 2005;129(1):113–21.
23. Bellentani S, Marino M. Epidemiology and natural history of non-alcoholic fatty liver disease (NAFLD). Ann Hepatol. 2009;8 Suppl 1:S4–8.
24. Browning JD, Szczepaniak LS, et al. Prevalence of hepatic steatosis in an urban population in the United States: impact of ethnicity. Hepatology. 2004;40(6):1387–95.
25. Charlton M. Nonalcoholic fatty liver disease: a review of current understanding and future impact. Clin Gastroenterol Hepatol. 2004;2(12):1048–58.
26. Fraser A, Longnecker MP, et al. Prevalence of elevated alanine aminotransferase among US adolescents and associated factors: NHANES 1999-2004. Gastroenterology. 2007;133(6):1814–20.
27. Lam B, Younossi ZM. Treatment options for nonalcoholic fatty liver disease. Therap Adv Gastroenterol. 2010;3(2):121–37.
28. Amarapurkar D, Kamani P, et al. Prevalence of non-alcoholic fatty liver disease: population based study. Ann Hepatol. 2007;6(3):161–3.
29. Bedogni G, Miglioli L, et al. Prevalence of and risk factors for nonalcoholic fatty liver disease: the Dionysos nutrition and liver study. Hepatology. 2005;42(1):44–52.
30. Zhou Y, Zheng S, et al. The interruption of the PDGF and EGF signaling pathways by curcumin stimulates gene expression of PPARgamma in rat activated hepatic stellate cell in vitro. Lab Invest. 2007;87(5):488–98.
31. Zhou YJ, Li YY, et al. Prevalence of fatty liver disease and its risk factors in the population of South China. World J Gastroenterol. 2007;13(47):6419–24.
32. Higuchi H, Gores GJ. Mechanisms of liver injury: an overview. Curr Mol Med. 2003;3(6): 483–90.
33. Day CP, James OF. Steatohepatitis: a tale of two "hits"? Gastroenterology. 1998;114(4):842–5.
34. Ip E, Farrell GC, et al. Central role of PPARalpha-dependent hepatic lipid turnover in dietary steatohepatitis in mice. Hepatology. 2003;38(1):123–32.
35. Reddy JK, Rao MS. Lipid metabolism and liver inflammation. II. Fatty liver disease and fatty acid oxidation. Am J Physiol Gastrointest Liver Physiol. 2006;290(5):G852–8.
36. Cai D, Yuan M, et al. Local and systemic insulin resistance resulting from hepatic activation of IKK-beta and NF-kappaB. Nat Med. 2005;11(2):183–90.
37. Braun S, Bitton-Worms K, et al. The link between the metabolic syndrome and cancer. Int J Biol Sci. 2011;7(7):1003–15.
38. Pollak M. Insulin and insulin-like growth factor signalling in neoplasia. Nat Rev Cancer. 2008;8(12):915–28.

39. Hursting SD, Smith SM, et al. Calories and cancer: the role of insulin-like growth factor-1. In: Leroith D, editor. The IGF system and cancer. New York: Springer; 2011. p. 231–43.
40. Wong KK, Engelman JA, et al. Targeting the PI3K signaling pathway in cancer. Curr Opin Genet Dev. 2010;20(1):87–90.
41. Memmott RM, Dennis PA. Akt-dependent and -independent mechanisms of mTOR regulation in cancer. Cell Signal. 2009;21(5):656–64.
42. Lindsley JE, Rutter J. Nutrient sensing and metabolic decisions. Comp Biochem Physiol B Biochem Mol Biol. 2004;139(4):543–59.
43. Moore T, Beltran L, et al. Dietary energy balance modulates signaling through the Akt/mammalian target of rapamycin pathways in multiple epithelial tissues. Cancer Prev Res (Phila). 2008;1(1):65–76.
44. De Angel RE, Conti CJ, et al. The enhancing effects of obesity on mammary tumor growth and Akt/mTOR pathway activation persist after weight loss and are reversed by RAD001. Mol Carcinog. 2013;52(6):446–58.
45. Nogueira LM, Dunlap SM, Ford NA, Hursting SD. Calorie restriction and rapamycin inhibit MMTV-Wnt-1 mammary tumor growth in a mouse model of postmenopausal obesity. Endocr Relat Cancer. 2012;19(1):57–68.
46. Anisimov VN. Metformin for aging and cancer prevention. Aging (Albany NY). 2010;2(11): 760–74.
47. Athar M, Kopelovich L. Rapamycin and mTORC1 inhibition in the mouse: skin cancer prevention. Cancer Prev Res (Phila). 2011;4(7):957–61.
48. Chaudhary SC, Kurundkar D, et al. Metformin, an antidiabetic agent reduces growth of cutaneous squamous cell carcinoma by targeting mTOR signaling pathway. Photochem Photobiol. 2012;88(5):1149–56.
49. Checkley LA, Rho O, et al. Rapamycin is a potent inhibitor of skin tumor promotion by 12-O-tetradecanoylphorbol-13-acetate. Cancer Prev Res (Phila). 2011;4(7):1011–20.
50. Tomimoto A, Endo H, et al. Metformin suppresses intestinal polyp growth in ApcMin/+ mice. Cancer Sci. 2008;99(11):2136–41.
51. Gautron L, Elmquist JK. Sixteen years and counting: an update on leptin in energy balance. J Clin Invest. 2011;121(6):2087–93.
52. Villanueva EC, Myers Jr MG. Leptin receptor signaling and the regulation of mammalian physiology. Int J Obes (Lond). 2008;32 Suppl 7:S8–12.
53. Vaiopoulos AG, Marinou K, et al. The role of adiponectin in human vascular physiology. Int J Cardiol. 2012;155(2):188–93.
54. Barb D, Williams CJ, et al. Adiponectin in relation to malignancies: a review of existing basic research and clinical evidence. Am J Clin Nutr. 2007;86(3):s858–66.
55. Stofkova A. Leptin and adiponectin: from energy and metabolic dysbalance to inflammation and autoimmunity. Endocr Regul. 2009;43(4):157–68.
56. Fenton JI, Hord NG, et al. Leptin, insulin-like growth factor-1, and insulin-like growth factor-2 are mitogens in ApcMin/+ but not Apc+/+ colonic epithelial cell lines. Cancer Epidemiol Biomarkers Prev. 2005;14(7):1646–52.
57. Stattin P, Lukanova A, et al. Obesity and colon cancer: does leptin provide a link? Int J Cancer. 2004;109(1):149–52.
58. Wu MH, Chou YC, et al. Circulating levels of leptin, adiposity and breast cancer risk. Br J Cancer. 2009;100(4):578–82.
59. Grossmann ME, Nkhata KJ, et al. Effects of adiponectin on breast cancer cell growth and signaling. Br J Cancer. 2008;98(2):370–9.
60. Rzepka-Gorska I, Bedner R, et al. Serum adiponectin in relation to endometrial cancer and endometrial hyperplasia with atypia in obese women. Eur J Gynaecol Oncol. 2008;29(6): 594–7.
61. Tian YF, Chu CH, et al. Anthropometric measures, plasma adiponectin, and breast cancer risk. Endocr Relat Cancer. 2007;14(3):669–77.

62. Jung CH, Rhee EJ, et al. The relationship of adiponectin/leptin ratio with homeostasis model assessment insulin resistance index and metabolic syndrome in apparently healthy korean male adults. Korean Diabetes J. 2010;34(4):237–43.
63. Labruna G, Pasanisi F, et al. High leptin/adiponectin ratio and serum triglycerides are associated with an "at-risk" phenotype in young severely obese patients. Obesity (Silver Spring). 2011;19(7):1492–6.
64. Mirza S, Qu HQ, et al. Adiponectin/leptin ratio and metabolic syndrome in a Mexican American population. Clin Invest Med. 2011;34(5):E290.
65. Ashizawa N, Yahata T, et al. Serum leptin-adiponectin ratio and endometrial cancer risk in postmenopausal female subjects. Gynecol Oncol. 2010;119(1):65–9.
66. Chen DC, Chung YF, et al. Serum adiponectin and leptin levels in Taiwanese breast cancer patients. Cancer Lett. 2006;237(1):109–14.
67. Cleary MP, Ray A, et al. Targeting the adiponectin:leptin ratio for postmenopausal breast cancer prevention. Front Biosci (Schol Ed). 2009;1:329–57.
68. Harvey AE, Lashinger LM, et al. The growing challenge of obesity and cancer: an inflammatory issue. Ann N Y Acad Sci. 2011;1229:45–52.
69. Olefsky JM, Glass CK. Macrophages, inflammation, and insulin resistance. Annu Rev Physiol. 2010;72:219–46.
70. Subbaramaiah K, Howe LR, et al. Obesity is associated with inflammation and elevated aromatase expression in the mouse mammary gland. Cancer Prev Res (Phila). 2011;4(3):329–46.
71. O'Rourke RW. Inflammation in obesity-related diseases. Surgery. 2009;145(3):255–9.
72. Karin M. Nuclear factor-kappaB in cancer development and progression. Nature. 2006; 441(7092):431–6.
73. Renehan AG, Roberts DL, et al. Obesity and cancer: pathophysiological and biological mechanisms. Arch Physiol Biochem. 2008;114(1):71–83.
74. Virchow R. Die Krankenhasften Geschwulste; Berlin, Germany. Aetologie der neoplastichen Geschwelste/Pathogenie der neoplastischen Geschwulste. 1863;58.
75. Aggarwal BB, Gehlot P. Inflammation and cancer: how friendly is the relationship for cancer patients? Curr Opin Pharmacol. 2009;9(4):351–69.
76. Del Prete A, Allavena P, et al. Molecular pathways in cancer-related inflammation. Biochem Med (Zagreb). 2011;21(3):264–75.
77. Ono M. Molecular links between tumor angiogenesis and inflammation: inflammatory stimuli of macrophages and cancer cells as targets for therapeutic strategy. Cancer Sci. 2008;99(8): 1501–6.
78. Coussens LM, Werb Z. Inflammation and cancer. Nature. 2002;420(6917):860–7.
79. Foltz CJ, Fox JG, et al. Spontaneous inflammatory bowel disease in multiple mutant mouse lines: association with colonization by *Helicobacter hepaticus*. Helicobacter. 1998;3(2): 69–78.
80. Allavena P, Sica A, Garlanda C, Mantovani A. The Yin-Yang of tumor-associated macrophages in neoplastic progression and immune surveillance. Immunol Rev. 2008;222:155–61.
81. Koki A, Khan NK, et al. Cyclooxygenase-2 in human pathological disease. Adv Exp Med Biol. 2002;507:177–84.
82. Kundu JK, Surh YJ. Inflammation: gearing the journey to cancer. Mutat Res. 2008;659(1–2): 15–30.
83. Byrne AM, Bouchier-Hayes DJ, et al. Angiogenic and cell survival functions of vascular endothelial growth factor (VEGF). J Cell Mol Med. 2005;9(4):777–94.
84. Liu Y, Tamimi RM, et al. The association between vascular endothelial growth factor expression in invasive breast cancer and survival varies with intrinsic subtypes and use of adjuvant systemic therapy: results from the Nurses' Health Study. Breast Cancer Res Treat. 2011; 129(1):175–84.
85. Cao Y. Angiogenesis modulates adipogenesis and obesity. J Clin Invest. 2007;117(9):2362–8.
86. Renehan AG. Body fatness and bevacizumab-based therapy in metastatic colorectal cancer. Gut. 2010;59(3):289–90.

87. Simkens LH, Koopman M, et al. Influence of body mass index on outcome in advanced colorectal cancer patients receiving chemotherapy with or without targeted therapy. Eur J Cancer. 2011;47(17):2560–7.
88. Iwaki T, Urano T, et al. PAI-1, progress in understanding the clinical problem and its aetiology. Br J Haematol. 2012;157(3):291–8.
89. Muldowney 3rd JA, Chen Q, et al. Pentoxifylline lowers plasminogen activator inhibitor 1 levels in obese individuals: a pilot study. Angiology. 2012;63(6):429–34.
90. Skurk T, Hauner H. Obesity and impaired fibrinolysis: role of adipose production of plasminogen activator inhibitor-1. Int J Obes Relat Metab Disord. 2004;28(11):1357–64.

Chapter 2
Increased Adiposity and Colorectal Cancer

Charles Bellows and Herbert Tilg

Abstract It has been clinically well established in the last years that obesity and obesity-related disorders, such as nonalcoholic fatty liver diseases, metabolic syndrome, and type 2 diabetes, are characterized by an increased risk of developing colorectal polyps and colorectal cancer. All other things equal, this risk appears to be strongest with an increased waist circumference and waist-to-hip ratios compared to one's body mass index. In addition, obese men seem to be more at risk for colon cancer than obese women. Interestingly, among colon cancer patients, epidemiologic studies suggest that obesity is also associated with increased rates for death and disease relapse compared to patients of normal weight. The increased risk of cancer in the obese suggests that weight loss could reduce this risk or improve survival in individual diagnosed with colon cancer, but this remain to be fully investigated. Importantly, the associations between anthropometric measurements and cancer are very clinically relevant as it might affect our colon cancer screening strategies suggesting screening certain populations earlier than recommended for the general population. The pathophysiology behind this association is still not understood but might involve mediators derived from the adipose tissue. Obesity is commonly associated with adipose tissue inflammation, systemic inflammation, and release of numerous adipocytokines into the circulation. An imbalance in these fundamentally important adipose tissue-derived factors could contribute to disease manifestation beyond the adipose tissue including colorectal cancer. Cells other than adipocytes, including adipose stromal cells and monocytes accumulating in adipose tissue, have surfaced as

C. Bellows
Department of Surgery, Tulane University, New Orleans, LA, USA
e-mail: cbellows@tulane.edu

H. Tilg (✉)
Christian Doppler Research Laboratory for Gut Inflammation,
Medical University Innsbruck, Innsbruck, Austria
e-mail: herbert.tilg@i-med.ac.at

M.G. Kolonin (ed.), *Adipose Tissue and Cancer*, DOI 10.1007/978-1-4614-7660-3_2,

important contributors to the adipocytokine pool. Finally, recently discovered trafficking of adipose stromal cells suggests paracrine adipocytokine signaling in tumor microenvironment.

2.1 Introduction

The incidence of obesity and associated disorders has risen dramatically worldwide in the last two decades. Obese individuals exhibit an increased risk of developing diseases such as atherosclerosis, type 2 diabetes, nonalcoholic fatty liver disease (NAFLD), and various cancers [1]. Epidemiological studies indicate that obesity represents a significant risk factor for the development of various cancers such as prostate and breast cancer, leading cancers in the Western world. An impressive body of evidence, however, also indicates that the risk of colorectal adenoma, and cancer (CRC) is increased in subjects with obesity and related metabolic syndrome [2, 3]. A recent French study suggests that obesity and weight gain are associated with early colorectal carcinogenesis in women, especially regarding the distal colon [4]. Colorectal cancer is the second leading cancer death in the Western world and its death rate correlates with body mass index [5]. Interestingly, the distribution of body fat appears to be an important factor, with abdominal obesity, defined by waist circumference or HR, appearing to be more predictive of colorectal cancer-specific survival than overall obesity (high BMI). Most sporadic CRCs arise from adenomatous polyps and detection and removal of these polyps is recognized as a highly effective preventive strategy [6]. Recent CRC screening studies suggest that obesity and an increased body mass index are a significant additional risk factor for the development of colonic polyps with evidence that advanced adenomas arise in men almost a decade earlier than in women [7]. In this chapter, we evaluate the association of weight and other anthropometric variables with the risk, mortality, and recurrence rate of colon cancer among individuals pre- and postdiagnosis as well as elaborate on the role of physical activity and potentially underlying mechanisms, on the association of colonic polyps and CRC with obesity-associated diseases such as NAFLD.

2.2 Anthropometric Measurements and Colon Cancer

There is convincing evidence that excess weight (obesity) and distribution of body fat can increase not only the risk but also the mortality and recurrence rate of colon cancer. Exactly how obesity is linked to colon cancer is open to debate. Currently, several epidemiologic studies have examined the relationship between anthropometric measurements [body mass index (BMI), waist circumference (WC), and waist-to-hip ratio (WHR)] and colorectal cancer with diverse finding. This may stem from inconsistency in the way BMI and anthropometric measurements were collected in these studies (i.e., self-reported, directly measured, or obtained retrospectively from medical records) leading to possible measurement errors or the fact the anthropometric measurement were considered at different time points with respect to the colorectal cancer diagnosis (i.e., before diagnosis, near the time of diagnosis, after diagnosis, or in different combinations).

Despite these inconsistencies, several prospective cohort investigations have still reported a strong positive association between BMI and the risk of colorectal cancer [8–13]. Case–control studies have also shown that overweight and obesity are associated with a modest increase in risk of colon cancer, but the data is less consistent [8, 14–17]. Interestingly, the distribution of body fat appears to be a more important factor, with abdominal obesity, which can be measured by waist circumference, showing the strongest association with colon cancer risk [11]. It also appears that the colorectal cancer incidence of obese patients has gender-specific characteristics. Among men, a higher BMI is strongly associated with increased risk of colorectal cancer [8, 13, 18, 19]. An association between BMI and waist circumference with colon cancer risk is also seen in women, but it is weaker. However, the relationship between obesity and colon cancer risk in women is thought to vary by the stage of life. Indeed, when studies stratify results by age, an association between BMI and colon cancer risk is observed in cohorts of younger women. For example, a large cohort study found a positive association between BMI and colorectal cancer among women less than 55 years of age but not among older women [20], suggesting that the influence of obesity on colon cancer risk diminishes in aging women, probably related to menopausal status. Indeed, menopausal status appears to modify the relationship between BMI and colon cancer with a strong association between BMI and colon cancer risk seen in premenopausal but not postmenopausal women [21].

The relationship between anthropometric variables and associated cancer-specific endpoints, such as overall mortality, cancer specific survival, disease-free survival, and recurrence rate, has also been examined. The majority of studies adjusted outcomes for a number of confounding variables including age, gender, stage, degree of differentiation, and type of treatment, allowing researchers to untangle these complex relationships. Most studies investigating the influence of obesity on survival of colon cancer patients used anthropometric measurements at the time of cancer diagnosis or after the cancers were surgically removed. Some of these studies included participants enrolled in clinical trials of adjuvant chemotherapies. Such studies have several positive qualities: height and weight are accurately measured because they are used to calculate the chemotherapy dose, cancer recurrence and cause of death are investigated prospectively, cancer treatment is well documented, and groups are uniform in terms of stage and performance status at the time of diagnosis. Overall, most of these studies detected that obese patients with stage II or stage III colon cancer that receive surgery and chemotherapy had higher rates of cancer recurrence and mortality compared with normal-weight patients [22–24]. Although this difference is most evident overall for patients with BMI ≥ 35 kg/m^2, some controversies exist and not all studies have shown a significant association between BMI and overall survival, cancer recurrence, or death [25–27].

Other studies investigating the influence of obesity on survival of colon cancer patients used BMI and other anthropometric measurements before diagnosis. Although earlier studies found no correlation between the patients precancer BMI and overall CRC survival [28], many more recent studies have all supported the finding that being obese prior to being diagnosed with colon cancer increases your risk of dying from the disease [29–32]. One of the largest, the Iowa women's health study, followed over 37,000 postmenopausal women for 20 years, with BMI and waist–hip ratio (WHR) self-measured before colon cancer diagnosis [29]. Among the 3 % that developed colorectal cancer, greater precancer anthropometric measures as well as a BMI of less than 18.5 kg/m^2 predicted poorer all-cause survival duration. This study also found that a higher abdominal adiposity measured by WHR and waist circumference was significantly associated with an increased risk for death from colon cancer [29]. Another study from Australia of over 40,000 volunteers (age 40–69 years), followed for 8–12 years, found that precancer BMI was not significantly associated with CRC outcome [30]. However, increased waist circumference was associated with reduced overall survival and reduced disease-specific survival. Increasing percent body fat, measured by bioelectrical impedance, was also associated with a decreased disease-specific survival [30].

Noticeably, more and more studies are now demonstrating the location of body fat tissue is the best predictor of all-cause and colorectal cancer mortality than general obesity (i.e., BMI). Moon and colleagues investigated the prognostic significance of abdominal obesity and BMI in 161 resectable colorectal cancer patients [33]. The visceral fat area (VFA) and subcutaneous fat area (SFA) were measured on archived digitalized computed tomography scans taken prior to surgery. VFA/SFA ratio was calculated and patients were organized into percentiles according to the degree of proportional visceral adiposity, and into overweight and normal weight groups according to their preoperative BMI. BMI and visceral adiposity showed no influence on overall survival. However, patients with a high VFA/SFA ratio (>50 percentile) had significantly lower disease-free survival rates compared to patients with a low VFA/SFA ratio. In a multivariate analysis, the presence of lymph node metastasis and VFA/SFA ratio were the only significant prognostic factors for disease-free survival [33].

One of the most intriguing observations is that colon cancer survival may be less likely for patients who are not only very obese but also too thin at diagnosis [34]. In a multi-institutional series of over 1,000 patients from Iran, underweight (BMI < 18.5 kg/m^2) patients had worse overall survival compared to normal weight patients (BMI 18.6–24.9 kg/m^2), while overweight (BMI 25–29.9 kg/m^2) patients had improved survival [35]. Obesity was not significantly associated with survival in this population. The negative prognostic effects of being underweight (i.e., low BMI) and survival have also been reported in cohort studies of Asian [36] and American [37] patient who underwent surgery for colon cancer.

The prognostic value of obesity in patients with metastatic colorectal cancer remained poorly defined. Guiu and colleagues found that obese (high BMI and visceral fat area) metastatic CRC patients that received Bevacizumab-based treatments, a monoclonal antibody against vascular endothelial growth factor, had a statistically significant shorter time to disease progression, as well as shorter overall survival [28]. In contract, in another chemotherapy trial, no association between BMI and overall survival could be demonstrated in over 700 patients with advanced CRC receiving chemotherapy plus Bevacizumab [38]. Similarly, a retrospective study investigating the correlation between BMI and survival after liver resection for metastatic colorectal cancer found no association [39].

Taking into account the increasing obese population and the fact that these individuals are overrepresented among colon cancer patients, we need to understand the mechanisms of how body weight and the distribution of adipose tissue negatively impacts colon cancer incidence and survival and develop novel therapeutic options for prevention and treatment.

2.3 Colon Cancer Location

A different etiologic factor for cancers of the proximal and distal colon has been supported by several epidemiologic studies. Consequently, investigators examined the relationship of obesity and colon cancer risks according to different tumor locations. Interestingly, several case–control studies have shown a positive association between BMI and site-specific tumors of the colon in men, and they have generally reported a stronger risk for tumors of the distal colon than for tumors of the proximal colon [14, 15, 40]. This site-specific characteristic is supported by the results of a recent prospective cohort study [41] but not a meta-analysis [19]. Indeed, a systematic review and meta-analysis showed no difference in the association between BMI and cancer risk between the proximal and distal colon, for men or women, but the subsite analyses included few studies [19]. In contrast, a recent prospective cohort study of over one million Asian men and women followed for 7 years found that a higher BMI was associated with an elevated risk for distal colon cancer in men and with marginally elevated risk for proximal colon cancer in women [42]. The Netherlands Cohort Study reported that all indicators of body fat in men appeared to be most strongly associated with tumors of the distal colon than with tumors at other anatomic subsites. However, associations by subsite in women in this study were not as clear [43]. In contrast, an apparent positive association between BMI and distal colon tumors has been reported in prospective cohort studies of Swedish [20] and American [44] women. Interestingly, distal tumors have been shown to have higher rates of chromosomal instability, p53 mutation, and DNA aneuploidy that may confer a worse prognosis [45, 46].

2.4 Obesity, Physical Activity, and Colon Cancer

A lack of exercise is both a cause and consequence of obesity. Colorectal cancer has been one of the most extensively studied cancers in relation to physical activity. In recent years, an increasing number of studies have reported an inverse association between physical activity and the risk of colon cancer and colon adenomas, the precursor lesion detected and removed during colonoscopy. These associations are not only believably by several biological mechanisms, but also a recent meta-analysis of 20 studies on physical activity and colon adenoma supports the findings from earlier studies that found that increased physical activity to be associated with a significant reduction in colon polyp risk [47]. Importantly, this analysis estimated the risk reduction associated with the physical activity and found a significant 16 % risk reduction in colon polyp when comparing the most to the least physically active [47]. Although the risk reductions were similar for men and women, the association was stronger when analyses were limited to advanced or large polyps, with a risk reduction of 35 %.

Several case–control and cohort studies have also consistently shown that physical activity can decrease the risk of colon cancer. In a meta-analysis of 52 studies (24 case–control and 28 cohort studies) examining the link between physical activity and colon cancer, a significant 24 % reduced risk of colon cancer in people who were most active compared with the least was found [48]. This supports other reviews of the association between physical activity and colon cancer in the Asian and European populations [49, 50]. The ability to draw conclusions about the amount of physical activity necessary to achieve a risk reduction is limited since each study quantified activity differently with some including manual work while others concentrated on leisure time exercise such as going to the gym or running. Despite this, a recent example provides some information. In the US Nurses' Health Study, Wolin and colleagues [51] report a 23 % risk reduction when comparing the most to the least active women. The most active women expended more than 21.5 metabolic equivalent (MET) hours per week over energy expenditure at rest in leisure-time physical activity, whereas the least active expended less than 2 MET hours per week. These levels are equivalent to brisk walking for some 5–6 h per week in the most active and 0.5 h per week in the least active. No association has been consistently found for physical activity and rectal cancer, and the consensus is that one is unlikely to exist [52]. Taken together, these data demonstrating physical activity's role in colon cancer prevention suggests that physical activity has a role across the carcinogenic process. It is thought physical activity is beneficial because it reduces inflammation, encourages good gut transit, and reduces levels of growth hormones and insulin in the body, which can fuel tumors. One intriguing observations is that the colon cancer risks may drop among people who are physically active, regardless of their weight.

Physical activity also appears to affect disease outcome and recurrence after diagnosis and treatment with the greatest effect on colon cancer incidence [53]. Data from the Melbourne Collaborative Cohort Study showed that increased central adiposity and a lack of regular physical activity prior to the diagnosis of colorectal cancer is associated with poorer overall and disease-specific survival [30]. While others showed that recreational physical activity after the diagnosis of nonmetastatic colorectal cancer reduces the risk of colorectal cancer-specific and overall mortality [54]. In a similar large cohort study (668 nonmetastatic CRC male patients from the Health Professional Follow-up Study), an increase in MET hours/week was also significantly associated with improved colorectal cancer-specific and overall mortality [55]. Men who engaged in more than 27 MET hours per week of postdiagnosis physical activity had an adjusted hazard ratio for colorectal cancer-specific mortality of 0.47 (95 % CI 0.24–0.92) and an adjusted hazard ratio for overall mortality of 0.59 (95 % CI 0.41–0.86) compared with men who engaged in 3 or less MET hours per week of postdiagnosis physical activity. More evidence correlating postdiagnosis physical activity and CRC survival was provided by an NCI intergroup trial of 832 stage III colon cancer patients undergoing different chemotherapy regimens [56]. Six months after completing therapy, the study showed that those who exercised regularly—the equivalent of walking 6 or more hours a week at a pace of 2–2.9 miles per hour—were almost 50 % more likely to be alive and free of cancer than those who were less physically active. A subsequent study of patients from the same trial focused on the association of BMI and weight changes after treatment with cancer recurrence and survival [26]. Again, the positive effect of exercise did not depend on weight or BMI. Weight change (either loss or gain) during the time period between ongoing adjuvant therapy and 6 months after completion of therapy did not significantly influence cancer recurrence and/or mortality. In the previously mentioned Cancer Prevention Study-II Nutrition Cohort, BMI was recorded at a mean of 7 years before diagnosis and 18 months after the diagnosis of CRC in 2,303 patients [32]. Postdiagnosis BMI was not found to be associated with overall mortality, cancer-specific mortality, or cardiovascular disease-specific mortality. These studies have direct clinical implications, since diet and exercise can be modified and it has been shown that cancer patients are motivated and capable of adjusting their lifestyle and dietary habits [57]. However, it is too early to tell if losing weight after being diagnosed with colon cancer will decrease the mortality risk.

2.5 Adipose Progenitor Cells

A possible link between obesity and cancer, which has so far been overlooked, relates to White Adipose Tissue (WAT) as a source of progenitor cells that could promote tumorigenesis and cancer progression. For a number of cancers, including colon cancer, outcome after diagnosis is adversely affected by obesity, thus suggesting that adiposity not only increases cancer risk but also promotes progression of cancer. Tumor vascularization relies on de novo vasculogenesis from adult endothelial progenitor cells that have been shown to originate in the bone marrow [58–60]. Multipotent progenitors for pericytes and other vascular cells have been also characterized [61]. Accumulating evidence suggests that adult mesenchymal stem (stromal) cells (MSC) [62, 63] represent a population of cells that can home to tumors and influence their growth [64, 65]. A plethora of reports have demonstrated that MSC secrete proangiogenic factors upon tumor engraftment and serve as vascular progenitor cells in pathological tissues [66]. Importantly, upregulation of stromal cell circulation has been detected in peripheral blood of cancer patients [67].

While the dependence of tumor growth on bone marrow-derived progenitors is unequivocal [68], it was recently realized that circulating stromal progenitor cells from other organs also contribute to postnatal neovascularization [69].

In the course of the past few years, mesenchymal progenitor cells similar to bone marrow MSC have been found to reside in the stromal vascular fraction (SVF) of adipose tissue [70]. These adipose stromal cells, commonly termed adipose stem cells (ASC), have been determined to reside within the pericytic niche that supports microvasculature [71]. The multipotency of ASC has been demonstrated by their ability to differentiate into tissues of mesenchymal lineage such as bone, cartilage, and WAT [72]. Patient-derived ASC have been shown to infiltrate sites of injury and promote angiogenesis through secreting angiogenic growth factors with local tropic effect, as well as to differentiate into perivascular and possibly even endothelial cells [72, 73]. These observations make it logical to expect that multipotent cells from WAT, excessively abundant in obese patients, may be recruited as a major source of stromal and/or vascular precursors for growing tumors. In particular, it should be noted that colon cancer and other cancers epidemiologically found to be advanced by obesity tend to occur in organs adjacent to natural WAT depots. This is consistent with a possibility that local migration of vascular progenitors is particularly important for colon tumors in men, who feature predominantly abdominal adiposity. A strong support to this hypothesis is the recent reports demonstrating that ASC can systemically home to and engraft into tumors [74] and can facilitate tumor take and growth in mouse tumor models [75]. Notably, the quantities and properties of ASC and of the differentiating preadipocytes have been found to differ between visceral and subcutaneous fat depots [76]. Visceral (e.g., omental) adipose progenitors appear to have comparatively high self-renewal capacity, as in age abdominal WAT remains, while other depots are depleted [77]. The notion of visceral WAT being the predominant driver of cancer is consistent with its possible involvement as the major reservoir of ASC in the elderly. Recent identification of factors predominantly secreted by omental but not subcutaneous WAT stromal vascular cells, such as omentin [78], points to possible new mechanisms linking visceral adiposity, adipose progenitor cells, and cancer.

While the importance of increasing blood supply through vasculogenesis and angiogenesis in cancers development is unequivocal, the stage at which it becomes critical in CRC remains unclear. Most studies have reported the angiogenic switch to occur once the invasive carcinoma has been established. However, even in a premalignant stage, epithelial cells have increased proliferation (as a manifestation of the "field effect") and therefore would be expected to require increased blood supply. Angiogenesis has previously been shown as early as in small adenomatous polyp or even in the aberrant crypt foci (ACF) stage. Moreover, abnormalities in the microvasculature of the "transitional mucosa" (histologically normal appearing epithelium adjacent to a CRC) suggest that alterations in blood supply may precede macroscopic neoplastic lesions. Therefore, ASC should be considered as a candidate entity to account for the increased cancer rate in obese patients, in addition to possibly playing a role in the promotion of cancer progression. In experimental models, it has been shown that MSC can execute a carcinogenic effect on low-grade tumor cells and cause the cancer cells to increase their metastatic potency [79]. Because ASC were also reported to facilitate cancer initiation in mice [80], it is possible that under the influence of tumor microenvironment ASC may play the role of mesenchymal tumor cells (MTC) and contribute to the carcinogenetic process in CRC.

The discussed evidences and the potential mechanisms for the association between obesity and cancer suggest reduction of body weight as a possible approach to prevent cancer or impede cancer progression. In various animal studies, caloric restriction dramatically decreased spontaneous and carcinogen-induced tumor incidence, multiplicity, and size [81]. Consistent with these observations, mortality in patients with certain types of cancer was significantly reduced by gastric bypass surgery [82]. However, in the meta-analysis, IARC did not find the evidence for a lack of significant association of low BMI with reduced risk of CRC in humans. These controversial results indicate that new well-controlled clinical trials on obesity prevention and obesity treatment are necessary before therapeutic implications of WAT reduction on cancer predisposition are completely understood. One of the possibly important considerations is the number of adipocytes and the accompanying stromal/vascular cells in WAT increasing in obesity and remaining increased even upon subsequent weight loss, which occurs via adipocyte size reduction. The pool of ASC is likely to remain intact and could contribute to cancer onset or progression despite calorie restriction and reduced adiposity. Provided the large and highly dynamic endothelial surface present in adipose tissue, it should not be excluded that even committed endothelial and perivascular cells from WAT are mobilized and could be functionally recruited by tumors upon WAT remodeling. In light of these considerations, weight loss measures such as dieting, as well as surgical interventions, in obese cancer patients may further contribute to progenitor cell mobilization. Recruitment of these WAT-derived progenitors by tumors could account for the subsequent cancer progression acceleration [83, 84]. Therefore, approaches to the inactivation of WAT-derived progenitor cells, such as those based on targeted depletion of WAT endothelium [85], may benefit the safety of cancer and obesity treatments and are now being developed [86]. In the meanwhile, it is apparent that prevention of obesity is the safest and the most natural strategy to decrease the likelihood of cancer development.

2.6 Adipocytokines and Adipose Tissue Inflammation: Driving Forces of Colonic Carcinogenesis

2.6.1 Adipose Tissue: A "Sink" of Tumor-Promoting Mediators?

The adipose tissue, which was previously considered a "rather primitive" organ mainly there to store energy, has been identified as a highly active endocrine organ, which releases an enormous number of so-called adipocytokines, mediators regulating throughout the body important metabolic and biological functions [87]. Obesity is associated with a chronic inflammatory state characterized by abnormal cytokine, adipocytokine, and acute-phase protein production and enhanced activity of inflammatory signaling pathways. Studies from the last years have found that in this state of chronic inflammation, the adipose tissue is the main production site of inflammatory mediators that link metabolism with the immune system and vice versa [88]. Although currently there is still limited evidence, inflammatory/immune mediators released by various cell types from the adipose tissue, including adipocytes and macrophages, could be attractive candidates for linking obesity with cancer.

Adiponectin has been identified as a key factor released by the adipose tissue, which important function in health is underpinned by the fact that it circulates at very high concentrations and serum concentrations decrease in case of obesity [89]. Adiponectin is a pleiotropic molecule with immunomodulatory, anti-inflammatory, cardioprotective, insulin-sensitizing, and antiangiogenic effects [90]. Various cancers, including CRC, gastric, endometrial, prostate, and breast cancer, are associated with decreased and low circulating adiponectin levels [91–95]. Visceral obesity, a condition commonly correlated with hypoadiponectinemia, is an established risk factor for CRC [96, 97]. Visceral adipose tissue in itself might also act protumorigenic and it has been recently suggested that its high content of vascular endothelial growth factor and interleukin-6 (IL-6) might contribute to cancer development [98]. Serum adiponectin levels are potentially not only related to CRC risk but also to tumor grade as observed in a study assessing 104 patients with newly diagnosed CRC [99]. In an also recently presented case–cohort study investigating serum samples from 457 CRC cases, low levels of adiponectin and high concentrations of leptin, plasminogen activator inhibitor-1, and IL-6 were associated with an increased CRC risk [100]. Whereas it is still not understood how for example adiponectin might interfere with adenoma development, certain genetic variants in the adiponectin gene have been associated with CRC development [92, 101, 102]. Certain adiponectin isoforms might also be of importance as another case–control study consisting of 778 cases and 735 controls demonstrated an inverse correlation of total and high-molecular weight adiponectin serum levels with colorectal adenoma development [103]. One of the hypothesis could be that chronic inflammation observed in obesity could reflect as one of the driving forces in CRC development as it is well known that other diseases characterized by chronic inflammation such as type 2 diabetes or ulcerative colitis are also associated by an increased CRC risk [104, 105].

Adiponectin might have certain effects on the homeostasis of epithelial cell proliferation as the colonic epithelium showed an increase in proliferation in adiponectin-deficient mice fed a high-fat diet [106]. Adiponectin has also been demonstrated to inhibit CRC cell growth by activation of adenosine monophosphate-activated protein kinase (AMPK) and suppression of mammalian-target-of-rapamycin pathways [107]. A similar effect has been observed in another study where adiponectin demonstrated an anticarcinogenic effect in vitro by inhibiting the growth of colon cancer cells again through stimulating AMPK activity [108]. In a mouse tumor model, adiponectin substantially inhibited primary tumor growth in a caspase-dependent manner resulting in endothelial-cell apoptosis [109]. Not all studies revealed an anticarcinogenic effect for adiponectin as observed in adiponectin transgenic mice challenged with the carcinogen azoxymethane and a high-fat diet for 24 weeks [110]. In this study, higher adiponectin levels had no effect on colon tumor incidence, numbers, or size of tumors. In contrast, Saxena and colleagues by inducing colonic tumors in mice by repeated injections of dextran sodium sulfate (DSS) and 1,2-dimethylhydrazine achieved different results [111]. Adiponectin KO mice in their studies had more evidence of inflammation and presented not only higher numbers but also significantly larger colonic tumors suggesting an anticancer role for adiponectin in their model. A role for its antitumor effect was also nicely demonstrated when adiponectin KO mice were crossed with Apc(Min) mice [112]. Phosphorylation of AMPK decreased in intestinal epithelial cells in adiponectin KO/Apc(Min) mice and these mice exhibited a 3.4-fold higher tumor formation compared to adiponectin$^{+/+}$/Apc(Min) mice. Similar observations were made when Apc(Min) mice were treated with recombinant adiponectin [113].

Leptin had been the first identified adipocytokines [114]. Many studies have investigated the effects of leptin, another important and often dysregulated adipocytokine in obesity and NAFLD, on different cancer types in experimental cellular and animal models [115]. Most of the studies indicate that leptin can potentiate the growth of cancer cells (breast, esophageal, gastric, pancreatic, colorectal, prostate, ovarian, and lung carcinoma cell lines), whereas adiponectin seems to decrease cell proliferation. Endo and colleagues demonstrated that leptin is of major importance for CRC growth in obesity and acts also as a growth factor for CRC at stages subsequent to tumor initiation in colorectal carcinogenesis [116] speculating that inhibition of leptin could reflect an attractive treatment concept. At least two clinical studies have convincingly demonstrated an inverse relationship between adiponectin and leptin serum levels and CRC [100, 103]. Adiponectin may have an anticarcinogenic effect on the large intestine by interfering with leptin, whereas leptin could conversely exert a carcinogenic effect under conditions of lower adiponectin availability. Although the mechanistic interrelations of various adipocytokines such as adiponectin and leptin in models of CRC are far from clear, adipocytokines could indeed act as a link between obesity and several gastrointestinal cancers. Several other mechanisms, such as reactive oxygen species, hyperinsulinemia, and increased circulating concentrations of insulin-like growth factor-1 by activation of phosphatidylinositol 3-kinase/Akt and mitogen-activated protein kinase/p38 signaling pathways, may in addition contribute to the increased risk of CRC in adiposity.

2.6.2 *JNK: Another Pathway in Obesity-Related Colorectal Carcinoma?*

C-Jun N-terminal kinases (JNKs) were initially characterized by their activation in response to cell stress such as UV irradiation. JNKs have since been characterized to be involved in proliferation, apoptosis, motility, metabolism, and DNA repair. Dysregulated JNK signaling is now believed to contribute to many diseases involving neurodegeneration, chronic inflammation, metabolic syndrome, cancer, and ischemia/reperfusion injury [117]. Activation of JNK/c-Jun has been observed in many different cancers including CRC and increased expression, and activation of JNK could play an important role in progression of CRC [118]. Endo and colleagues recently addressed the important question whether JNK might be involved in early colorectal carcinogenesis initiated by a high-fat diet and by repeated administration of azoxymethane [119]. The metabolic challenge of the colonic epithelium by this high-fat diet led to the occurrence of aberrant crypt foci and enhanced cell proliferation in the colonic epithelium associated with increased epithelial JNK activity. More importantly, specific JNK inhibitors could suppress enhanced cellular proliferation. Therefore, a high-fat diet, which is an established risk factor for CRC [120], activates the JNK pathway leading to the generation of an "insulin-resistant" epithelial cell, which could be one of the tumor-initiating events. Certain diets therefore might have "tumor-promoting" potential, which is of importance for humans as obesity is correlated with the increased incidence of CRC. There might also exist a link between JNK and certain adipocytokines such as adiponectin, as adiponectin might be able to suppress JNK activity. Although not yet proven, decreased adiponectin levels as observed in case of obesity might fail to control JNK expression in various tissues including the epithelial cell throughout the gut [121, 122]. Large efforts have been made in the last years towards understanding the molecules linking obesity, inflammation, and cancer. It becomes more and more evident that adipose tissue-derived mediators such as adiponectin or leptin and certainly many others are impaired in obesity and related chronic inflammation and thereby might exert detrimental effects at certain target organs, e.g., the colon and affect epithelial cell regulation, tissue homeostasis and finally contribute to tumor development.

2.7 Nonalcoholic Fatty Liver Disease and Increased Colon Carcinogenesis

NAFLD has evolved in parallel to the obesity pandemic as the most prevalent liver disease worldwide. Whereas the fact that chronic liver inflammation as observed in nonalcoholic steatohepatitis (NASH) finally leads to the development of hepatocellular carcinoma is well accepted [123], its association with increased formation of adenomatous polyps and CRC has just recently been established [124, 125]. Hwang and colleagues reported in 2010 the first evidence for an association of NAFLD with an increased rate of colorectal adenomatous polyps [126]. In their study, a population of almost 3,000 participants was investigated via colonoscopy, abdominal ultrasonography, and liver tests. The prevalence of NAFLD was over 40 % in the adenomatous polyp group versus 30 % in the control group providing the first evidence that such an association might indeed exist. Wong and colleagues recently published an excellent study in GUT convincingly showing that colorectal neoplasms are more prevalent in NASH subjects [124]. In their cross-sectional study,

NAFLD patients were defined histologically and by proton-magnetic resonance spectroscopy. Importantly, NAFLD patients had both a significantly higher rate of colorectal adenomas and advanced neoplasms than healthy controls. In addition, 13/29 (45 %) of NAFLD patients with advanced neoplasm had right-sided colorectal carcinoma. Furthermore, the presence of inflammation, i.e., NASH was a clear risk factor for adenoma and carcinoma development highlighting the fact that inflammation in a certain organ, e.g., in the liver might favor carcinogenesis at distal sites, e.g., the colon. This study performed in a large, well-defined population clearly suggests that NASH patients are indeed a risk population for the development of CRC and therefore screening colonoscopy should be strongly recommended in these patients. Findings of this study are supported by another report from Datz and colleagues [125]. They also provided strong evidence that NAFLD patients exhibit an increased risk of not only developing adenomatous polyps but also CRC.

References

1. Berrington de Gonzalez A, Hartge P, Cerhan JR, et al. Body-mass index and mortality among 1.46 million white adults. N Engl J Med. 2010;363:2211–9.
2. Donohoe CL, Pidgeon GP, Lysaght J, et al. Obesity and gastrointestinal cancer. Br J Surg. 2010;97:628–42.
3. Renehan AG, Soerjomataram I, Tyson M, et al. Incident cancer burden attributable to excess body mass index in 30 European countries. Int J Cancer. 2010;126:692–702.
4. Morois S, Mesrine S, Josset M, et al. Anthropometric factors in adulthood and risk of colorectal adenomas: the French E3N-EPIC prospective cohort. Am J Epidemiol. 2010;172:1166–80.
5. Dehal A, Garrett T, Tedders SH, et al. Body mass index and death rate of colorectal cancer among a national cohort of U.S. adults. Nutr Cancer. 2011;63:1218–25.
6. Zauber AG, Winawer SJ, O'Brien MJ, et al. Colonoscopic polypectomy and long-term prevention of colorectal-cancer deaths. N Engl J Med. 2012;366:687–96.
7. Ferlitsch M, Reinhart K, Pramhas S, et al. Sex-specific prevalence of adenomas, advanced adenomas, and colorectal cancer in individuals undergoing screening colonoscopy. JAMA. 2011;306:1352–8.
8. Ning Y, Wang L. Giovannucci ELA quantitative analysis of body mass index and colorectal cancer: findings from 56 observational studies. Obes Rev. 2010;11(1):19–30.
9. Chyou PH, Nomura AM, Stemmermann GN. A prospective study of weight, body mass index and other anthropometric measurements in relation to site-specific cancers. Int J Cancer. 1994;57(3):313–7.
10. Lin J, Zhang SM, Cook NR, Rexrode KM, Lee IM, Buring JE. Body mass index and risk of colorectal cancer in women (United States). Cancer Causes Control. 2004;15(6):581–9.
11. Moore LL, Bradlee ML, Singer MR, Splansky GL, Proctor MH, Ellison RC, et al. BMI and waist circumference as predictors of lifetime colon cancer risk in Framingham Study adults. Int J Obes Relat Metab Disord. 2004;28:559–67.
12. Ford ES. Body mass index and colon cancer in a national sample of adult US men and women. Am J Epidemiol. 1999;150:390–8.
13. Murphy TK, Calle EE, Rodriguez C, Kahn HS, Thun MJ. Body mass index and colon cancer mortality in a large prospective study. Am J Epidemiol. 2000;152:847–54.
14. Le Marchand L, Wilkens LR, Mi MP. Obesity in youth and middle age and risk of colorectal cancer in men. Cancer Causes Control. 1992;3:349–54.

15. Caan BJ, Coates AO, Slattery ML, Potter JD, Quesenberry Jr CP, Edwards SM. Body size and the risk of colon cancer in a large case-control study. Int J Obes Relat Metab Disord. 1998;22:178–84.
16. Dietz AT, Newcomb PA, Marcus PM, Storer BE. The association of body size and large bowel cancer risk in Wisconsin (United States) women. Cancer Causes Control. 1995; 6:30–6.
17. Russo A, Franceschi S, La Vecchia C, et al. Body size and colorectal-cancer risk. Int J Cancer. 1998;78:161–5.
18. Larsson SC, Wolk A. Obesity and colon and rectal cancer risk: a meta-analysis of prospective studies. Am J Clin Nutr. 2007;86:556–65.
19. Harriss DJ, Atkinson G, George K, et al. Lifestyle factors and colorectal cancer risk (1): systematic review and meta-analysis of associations with body mass index. Colorectal Dis. 2009;11:547–63.
20. Terry P, Giovannucci E, Bergkvist L, et al. Body weight and colorectal cancer risk in a cohort of Swedish women: relation varies by age and cancer site. Br J Cancer. 2001;85:346–9.
21. Terry PD, Miller AB, Rohan TE. Obesity and colorectal cancer risk in women. Gut. 2002;51:191–4.
22. Meyerhardt JA et al. Influence of body mass index on outcomes and treatment-related toxicity in patients with colon carcinoma. Cancer. 2003;98:484–95.
23. Dignam JJ, Polite BN, Yothers G, Raich P, Colangelo L, O'Connell MJ, et al. Body mass index and outcomes in patients who receive adjuvant chemotherapy for colon cancer. J Natl Cancer Inst. 2006;98:1647–54.
24. Sinicrope FA et al. Obesity is an independent prognostic variable in colon cancer survivors. Clin Cancer Res. 2010;16:1884–93.
25. Healy LA et al. Impact of obesity on surgical and oncological outcomes in the management of colorectal cancer. Int J Colorectal Dis. 2010;25:1293–9.
26. Meyerhardt JA et al. Impact of body mass index and weight change after treatment on cancer recurrence and survival in patients with stage III colon cancer: findings from Cancer and Leukemia Group B 89803. J Clin Oncol. 2008;26:4109–15.
27. Roxburgh CS et al. Relationship between preoperative comorbidity, systemic inflammatory response, and survival in patients undergoing curative resection for colorectal cancer. Ann Surg Oncol. 2011;18:997–1005.
28. Slattery ML et al. Diet and survival of patients with colon cancer in Utah: is there an association? Int J Epidemiol. 1989;18:792–7.
29. Prizment AE, Flood A, Anderson KE, Folsom AR. Survival of women with colon cancer in relation to precancer anthropometric characteristics: the Iowa Women's Health Study. Cancer Epidemiol Biomarkers Prev. 2010;19:2229–37.
30. Haydon AM, Macinnis RJ, English DR, Giles GG. Effect of physical activity and body size on survival after diagnosis with colorectal cancer. Gut. 2006;55:62–7.
31. Park SM et al. Impact of prediagnosis smoking, alcohol, obesity, and insulin resistance on survival in male cancer patients: National Health Insurance Corporation Study. J Clin Oncol. 2006;24:5017–24.
32. Campbell PT et al. Impact of body mass index on survival after colorectal cancer diagnosis: the Cancer Prevention Study-II Nutrition Cohort. J Clin Oncol. 2011;30(1):42–52.
33. Moon HG et al. Visceral obesity may affect oncologic outcome in patients with colorectal cancer. Ann Surg Oncol. 2008;15:1918–22.
34. Doria-Rose VP et al. Body mass index and the risk of death following the diagnosis of colorectal cancer in postmenopausal women (United States). Cancer Causes Control. 2006;17:63–70.
35. Asghari-Jafarabadi M et al. Site-specific evaluation of prognostic factors on survival in Iranian colorectal cancer patients: a competing risks survival analysis. Asian Pac J Cancer Prev. 2009;10:815–21.
36. Shibakita M et al. Body mass index influences long-term outcome in patients with colorectal cancer. Hepatogastroenterology. 2010;57:62–9.

37. Hines RB et al. Effect of comorbidity and body mass index on the survival of African-American and Caucasian patients with colon cancer. Cancer. 2009;115:5798–806.
38. Simkens LH et al. Influence of body mass index on outcome in advanced colorectal cancer patients receiving chemotherapy with or without targeted therapy. Eur J Cancer. 2011;47: 2560–7.
39. Pathak S et al. Hepatic steatosis, body mass index and long term outcome in patients undergoing hepatectomy for colorectal liver metastases. Eur J Surg Oncol. 2010;36:52–7.
40. West DW, Slattery ML, Robison LM, et al. Dietary intake and colon cancer: sex- and anatomic site-specific associations. Am J Epidemiol. 1989;130:883–94.
41. Laake I, Thune I, Selmer R, et al. A prospective study of body mass index, weight change, and risk of cancer in the proximal and distal colon. Cancer Epidemiol Biomarkers Prev. 2010;19:1511–22.
42. Shin A, Joo J, Bak J, Yang HR, Kim J, Park S, et al. Site-specific risk factors for colorectal cancer in a Korean population. PLoS One. 2011;6:e23196.
43. Hughes LA, Simons CC, van den Brandt PA, Goldbohm RA, van Engeland M, Weijenberg MP. Body size and colorectal cancer risk after 16.3 years of follow-up: an analysis from the Netherlands Cohort Study. Am J Epidemiol. 2011;174:1127–39.
44. Oxentenko AS, Bardia A, Vierkant RA, et al. Body size and incident colorectal cancer: a prospective study of older women. Cancer Prev Res (Phila). 2010;3:1608–20.
45. Offerhaus GJ, De Feyter EP, Cornelisse CJ, et al. The relationship of DNA aneuploidy to molecular genetic alterations in colorectal carcinoma. Gastroenterology. 1992;102:1612–9.
46. Bell SM, Scott N, Cross D, et al. Prognostic value of p53 over-expression and c-Ki-ras gene mutations in colorectal cancer. Gastroenterology. 1993;104:57–64.
47. Wolin KY, Yan Y, Colditz GA. Physical activity and risk of colon adenoma: a meta-analysis. Br J Cancer. 2011;104:882–5.
48. Wolin KY, Yan Y, Colditz GA, Lee IM. Physical activity and colon cancer prevention: meta-analysis. Br J Cancer. 2009;100:611–6.
49. Pham NM, Mizoue T, Tanaka K, Tsuji I, Tamakoshi A, Matsuo K, et al. Physical activity and colorectal cancer risk: an evaluation based on a systematic review of epidemiologic evidence among the Japanese population. Jpn J Clin Oncol. 2012;42:2–13.
50. Nilsen TIL, PlR R, Petersen H, Gunnell D, Vatten LJ. Recreational physical activity and cancer risk in subsites of the colon (the Nord-Trøndelag Health Study). Cancer Epidemiol Biomarkers Prev. 2008;17:183–8.
51. Wolin KY, Lee IM, Colditz GA, Glynn RJ, Fuchs C, Giovannucci E. Leisure-time physical activity patterns and risk of colon cancer in women. Int J Cancer. 2007;121(12):2776–81.
52. International Agency for Research on Cancer WHO. IARC Handbooks of cancer prevention: weight control and physical activity, vol. 6. Lyon: International Agency for Research on Cancer; 2002.
53. Denlinger CS, Engstrom PF. Colorectal cancer survivorship: movement matters. Cancer Prev Res (Phila). 2011;4:502–11.
54. Meyerhardt JA et al. Physical activity and survival after colorectal cancer diagnosis. J Clin Oncol. 2006;24:3527–34.
55. Meyerhardt JA et al. Physical activity and male colorectal cancer survival. Arch Intern Med. 2009;169:2102–8.
56. Meyerhardt JA et al. Impact of physical activity on cancer recurrence and survival in patients with stage III colon cancer: findings from CALGB 89803. J Clin Oncol. 2006;24:3535–41.
57. Satia JA, Campbell MK, Galanko JA, James A, Carr C, Sandler RS. Longitudinal changes in lifestyle behaviors and health status in colon cancer survivors. Cancer Epidemiol Biomarkers Prev. 2004;13:1022–31.
58. Kaplan RN, Riba RD, Zacharoulis S, Bramley AH, Vincent L, Costa C, et al. VEGFR1-positive haematopoietic bone marrow progenitors initiate the pre-metastatic niche. Nature. 2005;438:820–7.
59. Carmeliet P, Luttun A. The emerging role of the bone marrow-derived stem cells in (therapeutic) angiogenesis. Thromb Haemost. 2001;86:289–97.

60. Shaked Y, Ciarrocchi A, Franco M, Lee CR, Man S, Cheung AM, et al. Therapy-induced acute recruitment of circulating endothelial progenitor cells to tumors. Science. 2006;313: 1785–7.
61. Bergers G, Song S. The role of pericytes in blood-vessel formation and maintenance. Neuro Oncol. 2005;7:452–64.
62. Nilsson SK, Simmons PJ. Transplantable stem cells: home to specific niches. Curr Opin Hematol. 2004;11:102–6.
63. Prockop DJ. Marrow stromal cells as stem cells for nonhematopoietic tissues. Science. 1997;276:71–4.
64. Hung SC, Deng WP, Yang WK, Liu RS, Lee CC, Su TC, et al. Mesenchymal stem cell targeting of microscopic tumors and tumor stroma development monitored by noninvasive in vivo positron emission tomography imaging. Clin Cancer Res. 2005;11:7749–56.
65. Studeny M, Marini FC, Dembinski JL, Zompetta C, Cabreira-Hansen M, Bekele BN, et al. Mesenchymal stem cells: potential precursors for tumor stroma and targeted-delivery vehicles for anticancer agents. J Natl Cancer Inst. 2004;96:1593–603.
66. Liu JW, Dunoyer-Geindre S, Serre-Beinier V, Mai G, Lambert JF, Fish RJ, et al. Characterization of endothelial-like cells derived from human mesenchymal stem cells. J Thromb Haemost. 2007;5:826–34.
67. Fernández M, Simon V, Herrera G, Cao C, Del Favero H, Minguell JJ. Detection of stromal cells in peripheral blood progenitor cell collections from breast cancer patients. Bone Marrow Transplant. 1997;20:265–71.
68. Gao D, Nolan DJ, Mellick AS, Bambino K, McDonnell K, Mittal V. Endothelial progenitor cells control the angiogenic switch in mouse lung metastasis. Science. 2008;319:195–8.
69. Aicher A, Rentsch M, Sasaki K, Ellwart JW, Fändrich F, Siebert R, et al. Nonbone marrow-derived circulating progenitor cells contribute to postnatal neovascularization following tissue ischemia. Circ Res. 2007;100:581–9.
70. Zuk PA, Zhu M, Mizuno H, Huang J, Futrell JW, Katz AJ, et al. Multilineage cells from human adipose tissue: implications for cell-based therapies. Tissue Eng. 2001;7:211–28.
71. Traktuev DO, Merfeld-Clauss S, Li J, Kolonin M, Arap W, Pasqualini R, et al. A population of multipotent CD34-positive adipose stromal cells share pericyte and mesenchymal surface markers, reside in a periendothelial location, and stabilize endothelial networks. Circ Res. 2008;102:77–85.
72. Gimble JM, Guilak F, Nuttall ME, Sathishkumar S, Vidal M, Bunnell BA. In vitro differentiation potential of mesenchymal stem cells. Transfus Med Hemother. 2008;35:228–38.
73. Rehman J, Traktuev D, Li J, Merfeld-Clauss S, Temm-Grove CJ, Bovenkerk JE, et al. Secretion of angiogenic and antiapoptotic factors by human adipose stromal cells. Circulation. 2004;109:1292–8.
74. Kucerova L, Altanerova V, Matuskova M, Tyciakova S, Altaner C. Adipose tissue-derived human mesenchymal stem cells mediated prodrug cancer gene therapy. Cancer Res. 2007;67: 6304–13.
75. Martin-Padura I, Gregato G, Marighetti P, Mancuso P, Calleri A, Corsini C, et al. The white adipose tissue used in lipotransfer procedures is a rich reservoir of CD34+ progenitors able to promote cancer progression. Cancer Res. 2012;72:325–34.
76. Van Harmelen V, Röhrig K, Hauner H. Comparison of proliferation and differentiation capacity of human adipocyte precursor cells from the omental and subcutaneous adipose tissue depot of obese subjects. Metabolism. 2004;53:632–7.
77. Cartwright MJ, Tchkonia T, Kirkland JL. Aging in adipocytes: potential impact of inherent, depot-specific mechanisms. Exp Gerontol. 2007;42:463–71.
78. Yang RZ, Lee MJ, Hu H, Pray J, Wu HB, Hansen BC, et al. Identification of omentin as a novel depot-specific adipokine in human adipose tissue: possible role in modulating insulin action. Am J Physiol Endocrinol Metab. 2006;290(6):E1253–61.
79. Karnoub AE, Dash AB, Vo AP, Sullivan A, Brooks MW, Bell GW, et al. Mesenchymal stem cells within tumour stroma promote breast cancer metastasis. Nature. 2007;449:557–63.

80. Cawthorn WP, Scheller EL, Macdougald OA. Adipose tissue stem cells: the great WAT hope. Trends Endocrinol Metab. 2012;23:270–7.
81. Hursting SD, Lavigne JA, Berrigan D, Perkins SN, Barrett JC. Calorie restriction, aging, and cancer prevention: mechanisms of action and applicability to humans. Annu Rev Med. 2003; 54:131–52.
82. Adams TD, Stroup AM, Gress RE, Adams KF, Calle EE, Smith SC, et al. Cancer incidence and mortality after gastric bypass surgery. Obesity (Silver Spring). 2009;17:796–802.
83. Zhang Y, Bellows CF, Kolonin MG. Adipose tissue-derived progenitor cells and cancer. World J Stem Cells. 2010;2:103–13.
84. Bellows CF, Zhang Y, Chen J, Frazier ML, Kolonin MG. Circulation of progenitor cells in obese and lean colorectal cancer patients. Cancer Epidemiol Biomarkers Prev. 2011;20: 2461–8.
85. Kolonin MG, Saha PK, Chan L, Pasqualini R, Arap W. Reversal of obesity by targeted ablation of adipose tissue. Nat Med. 2004;10:625–32.
86. Daquinag AC, Zhang Y, Kolonin MG. Vascular targeting of adipose tissue as an anti-obesity approach. Trends Pharmacol Sci. 2011;32:300–7.
87. Sun K, Kusminski CM, Scherer PE. Adipose tissue remodeling and obesity. J Clin Invest. 2011;121:2094–101.
88. Tilg H, Moschen AR. Adipocytokines: mediators linking adipose tissue, inflammation and immunity. Nat Rev Immunol. 2006;6:772–83.
89. Shetty S, Kusminski CM, Scherer PE. Adiponectin in health and disease: evaluation of adiponectin-targeted drug development strategies. Trends Pharmacol Sci. 2009;30:234–9.
90. Tilg H, Moschen AR. Role of adiponectin and PBEF/visfatin as regulators of inflammation: involvement in obesity-associated diseases. Clin Sci (Lond). 2008;114:275–88.
91. Wei EK, Giovannucci E, Fuchs CS, et al. Low plasma adiponectin levels and risk of colorectal cancer in men: a prospective study. J Natl Cancer Inst. 2005;97:1688–94.
92. Kaklamani VG, Wisinski KB, Sadim M, et al. Variants of the adiponectin (ADIPOQ) and adiponectin receptor 1 (ADIPOR1) genes and colorectal cancer risk. JAMA. 2008;300: 1523–31.
93. Ishikawa M, Kitayama J, Kazama S, et al. Plasma adiponectin and gastric cancer. Clin Cancer Res. 2005;11:466–72.
94. Pinthus JH, Kleinmann N, Tisdale B, et al. Lower plasma adiponectin levels are associated with larger tumor size and metastasis in clear-cell carcinoma of the kidney. Eur Urol. 2008;54:866–73.
95. Cust AE, Stocks T, Lukanova A, et al. The influence of overweight and insulin resistance on breast cancer risk and tumour stage at diagnosis: a prospective study. Breast Cancer Res Treat. 2009;113:567–76.
96. An W, Bai Y, Deng SX, et al. Adiponectin levels in patients with colorectal cancer and adenoma: a meta-analysis. Eur J Cancer Prev. 2012;21:126–33.
97. Aleksandrova K, Boeing H, Jenab M, et al. Total and high-molecular weight adiponectin and risk of colorectal cancer: the European Prospective Investigation into Cancer and Nutrition Study. Carcinogenesis. 2012;33(2):1211–8.
98. Lysaght J, van der Stok EP, Allott EH, et al. Pro-inflammatory and tumour proliferative properties of excess visceral adipose tissue. Cancer Lett. 2011;312:62–72.
99. Gialamas SP, Petridou ET, Tseleni-Balafouta S, et al. Serum adiponectin levels and tissue expression of adiponectin receptors are associated with risk, stage, and grade of colorectal cancer. Metabolism. 2011;60:1530–8.
100. Ho GY, Wang T, Gunter MJ, et al. Adipokines linking obesity with colorectal cancer risk in postmenopausal women. Cancer Res. 2012;72(12):3029–37.
101. Pechlivanis S, Bermejo JL, Pardini B, et al. Genetic variation in adipokine genes and risk of colorectal cancer. Eur J Endocrinol. 2009;160:933–40.
102. Liu L, Zhong R, Wei S, et al. Interactions between genetic variants in the adiponectin, adiponectin receptor 1 and environmental factors on the risk of colorectal cancer. PLoS One. 2011;6:e27301.

103. Yamaji T, Iwasaki M, Sasazuki S, et al. Interaction between adiponectin and leptin influences the risk of colorectal adenoma. Cancer Res. 2010;70:5430–7.
104. Larsson SC, Orsini N, Wolk A. Diabetes mellitus and risk of colorectal cancer: a meta-analysis. J Natl Cancer Inst. 2005;97:1679–87.
105. Danese S, Fiocchi C. Ulcerative colitis. N Engl J Med. 2011;365:1713–25.
106. Fujisawa T, Endo H, Tomimoto A, et al. Adiponectin suppresses colorectal carcinogenesis under the high-fat diet condition. Gut. 2008;57:1531–8.
107. Sugiyama M, Takahashi H, Hosono K, et al. Adiponectin inhibits colorectal cancer cell growth through the AMPK/mTOR pathway. Int J Oncol. 2009;34:339–44.
108. Kim AY, Lee YS, Kim KH, et al. Adiponectin represses colon cancer cell proliferation via AdipoR1- and -R2-mediated AMPK activation. Mol Endocrinol. 2010;24:1441–52.
109. Brakenhielm E, Veitonmaki N, Cao R, et al. Adiponectin-induced antiangiogenesis and antitumor activity involve caspase-mediated endothelial cell apoptosis. Proc Natl Acad Sci USA. 2004;101:2476–81.
110. Ealey KN, Archer MC. Elevated circulating adiponectin and elevated insulin sensitivity in adiponectin transgenic mice are not associated with reduced susceptibility to colon carcinogenesis. Int J Cancer. 2009;124:2226–30.
111. Saxena A, Chumanevich A, Fletcher E, et al. Adiponectin deficiency: role in chronic inflammation induced colon cancer. Biochim Biophys Acta. 2012;1822:527–36.
112. Mutoh M, Teraoka N, Takasu S, et al. Loss of adiponectin promotes intestinal carcinogenesis in Min and wild-type mice. Gastroenterology. 2011;140:2000-8, 8 e1-2.
113. Otani K, Kitayama J, Yasuda K, et al. Adiponectin suppresses tumorigenesis in Apc(Min)(/+) mice. Cancer Lett. 2010;288:177–82.
114. La Cava A, Matarese G. The weight of leptin in immunity. Nat Rev Immunol. 2004;4:371–9.
115. Procaccini C, Galgani M, De Rosa V, et al. Leptin: the prototypic adipocytokine and its role in NAFLD. Curr Pharm Des. 2010;16:1902–12.
116. Endo H, Hosono K, Uchiyama T, et al. Leptin acts as a growth factor for colorectal tumours at stages subsequent to tumour initiation in murine colon carcinogenesis. Gut. 2011;60:1363–71.
117. Hirosumi J, Tuncman G, Chang L, et al. A central role for JNK in obesity and insulin resistance. Nature. 2002;420:333–6.
118. Fang JY, Richardson BC. The MAPK signalling pathways and colorectal cancer. Lancet Oncol. 2005;6:322–7.
119. Endo H, Hosono K, Fujisawa T, et al. Involvement of JNK pathway in the promotion of the early stage of colorectal carcinogenesis under high-fat dietary conditions. Gut. 2009;58:1637–43.
120. Tsugane S, Inoue M. Insulin resistance and cancer: epidemiological evidence. Cancer Sci. 2010;101:1073–9.
121. Jung TW, Lee YJ, Lee MW, et al. Full-length adiponectin protects hepatocytes from palmitate-induced apoptosis via inhibition of c-Jun NH2 terminal kinase. FEBS J. 2009;276:2278–84.
122. Kim KY, Kim JK, Jeon JH, et al. c-Jun N-terminal kinase is involved in the suppression of adiponectin expression by TNF-alpha in 3T3-L1 adipocytes. Biochem Biophys Res Commun. 2005;327:460–7.
123. Stickel F, Hellerbrand C. Non-alcoholic fatty liver disease as a risk factor for hepatocellular carcinoma: mechanisms and implications. Gut. 2010;59:1303–7.
124. Wong VW, Wong GL, Tsang SW, et al. High prevalence of colorectal neoplasm in patients with non-alcoholic steatohepatitis. Gut. 2011;60:829–36.
125. Stadlmayr A, Aigner E, Steger B, et al. Nonalcoholic fatty liver disease: an independent risk factor for colorectal neoplasia. J Intern Med. 2011;270:41–9.
126. Hwang ST, Cho YK, Park JH, et al. Relationship of non-alcoholic fatty liver disease to colorectal adenomatous polyps. J Gastroenterol Hepatol. 2010;25:562–7.

Chapter 3
Adiposity and Diabetes in Breast and Prostate Cancer

Linda Vona-Davis and David P. Rose

Abstract Obesity and type 2 diabetes are two of the most common potentially life-threatening diseases among men and women in the developed and developing countries of the world. Both diseases share common factors in the risk and prognosis of breast and prostate cancers. Adiposity, together with type 2 diabetes, promotes an aggressive and metastatic phenotype via endocrine and paracrine mechanisms of action by increasing the expression patterns of hormones and adipokines that are proposed to drive tumor grown. Understanding how the expansion of white adipose tissue, in addition to diabetes, alters the tumor microenvironment could provide links between obesity, metabolic diseases, and cancer progression. The purpose of this chapter is to examine the complex relationships between obesity and type 2 diabetes, and cancers of the breast and prostate, focusing on the biological mechanisms involved, and explaining how adipose tissue acts as a local source of steroid hormones, growth factors, and adipokines, which stimulate the growth cancer cells. Furthermore, the diametrically opposed effects of obesity, insulin resistance, and type 2 diabetes on premenopausal and postmenopausal carcinoma of the breast are discussed, as well as their potential influence on the development of the metastatic phenotype in both breast and prostate cancers.

3.1 Introduction

Carcinomas of the breast and prostate are two of the most commonly diagnosed cancers, and among the most frequent causes of cancer deaths in North America and Europe, and are showing rapidly increasing incidence and mortality rates in the

L. Vona-Davis, Ph.D. (✉) • D.P. Rose, M.D., Ph.D., D.Sc.
Department of Surgery and Breast Cancer Research Program, Mary Babb Randolph Cancer Center, West Virginia University Robert C. Byrd Health Sciences Center, P.O. Box 9238, Morgantown, WV 26506, USA
e-mail: lvdavis@hsc.wvu.edu

M.G. Kolonin (ed.), *Adipose Tissue and Cancer*, DOI 10.1007/978-1-4614-7660-3_3,

developing countries of the world. In the USA, breast cancer is the most frequently diagnosed cancer in women other than skin cancer and the second most common cause of cancer-related deaths; prostate cancer holds the same rankings in American men [1].

Both breast and prostate cancers are hormone-dependent tumors. Breast cancer occurs only infrequently in women who have undergone bilateral oophorectomy before the menopause for a reason other than cancer, whereas estrogen replacement therapy is associated with increased risk in postmenopausal women [2]. Similarly, carcinoma of the prostate does not occur in eunuchs, and these cancers exhibit androgen dependence and finasteride, a 5α-reductase inhibitor, which blocks biosynthesis of the potent androgen dihydrotestosterone, has been shown to reduce prostate cancer risk [3]. The expression of estrogen receptors (ER) by breast cancer epithelial cells is indicative of estrogen dependence, and ER-positive tumors also typically have a relatively good prognosis [4]. Androgen receptors (AR) have an important role in the biology of prostate cancer cells [5], but they have not achieved the clinical value of the breast cancer ER as a biomarker of hormone dependence or potential responsiveness to endocrine therapy.

Over the past 20 years or more, both cancers of the breast and prostate have been studied extensively in relation to two metabolic disorders, obesity and type 2 diabetes (T2D), which themselves occur in association with each other. The usual means of expressing adiposity is by way of the body mass index (BMI), which is calculated as body weight (kg)/height (m^2). The World Health Organization has defined the degrees of adiposity in terms of the BMI as follows: underweight, less than 18.5; normal, 18.5–24.9; overweight, 25.0–29.9; and obese, 30 or more kg/m^2. The prevalence of obesity is increasing in most developed and developing countries, and according to the National Health and Nutrition Examination Survey (NHANES), the age-adjusted prevalence of obesity in the USA for 2007–2008 was 35.5 % of adult women and 32.2 % of adult men [6]. What has been described as a worldwide pandemic of diabetes mellitus is also taking place; the NHANES survey for 2005–2006 showed that 7.7 % of the US population aged 20 years or older had previously diagnosed, and another 5.1 % previously unrecognized diabetes mellitus [7]. There are two forms of diabetes: type 1 is diagnosed most often in childhood and adolescence and results from an absolute deficiency of insulin that arises from autoimmune destruction of the pancreatic β-cells. Type 2 diabetes (T2D), which makes up about 95 % of all cases, typically occurs in middle age and in the elderly; it is a consequence of metabolic resistance to insulin action and is the form commonly seen in association with obesity.

The purpose of this chapter is to examine the complex relationships between obesity and T2D, and cancers of the breast and prostate. Our focus is on the biological mechanisms involved, with emphasis on the adipose tissue as an endocrine organ, as well as a local source of the steroid hormones and adipokines, which exert their effects on the cancer cells by paracrine mechanisms. The diametrically opposed effects of obesity, insulin resistance, and T2D on premenopausal and postmenopausal carcinoma of the breast are discussed, as well as their influence on the expression of the metastatic phenotype in both breast and prostate cancer.

3.2 Breast Cancer

Breast cancer is a malignant growth that originates in the milk ducts (ductal) or hollow glands (lobular) of the breast. It occurs in both men and women, although male breast cancer is rare. The most common type of breast cancer is invasive or infiltrating ductal carcinoma, when cancer cells invade nearby healthy breast tissue. When breast cancer becomes invasive, it has the potential to metastasis to distant sites. A small fraction of breast cancers are related to genetic mutations; however, the future of breast cancer prevention depends on understanding ways to lower risk and improve prognosis by controlling body weight and insulin levels.

3.3 Obesity and Breast Cancer Risk

3.3.1 Premenopause

We have reviewed previously the interaction between menopausal status and obesity in affecting breast cancer risk [8, 9]. In general, prospective studies, and most of case–control design, showed there to be an unaltered or reduced risk in obese premenopausal women and this was confirmed by two pooled analyses, which between them included the data from 27 studies carried out in 16 different countries [10, 11]. Overall, it was found that the inverse association occurred particularly in the younger women [11]. For example, one particularly well-designed case–control study, which was limited to American women aged 21–45 years, found a reduction in breast cancer risk related to a high BMI, but it was restricted to those who were interviewed when they were 35 years or younger [12]. Since 2000, a number of prospective studies have reported a (20–40 %) *decrease* in the risk of breast cancer in obese women compared with those considered of normal weight, but a recent report by Cecchini et al. [13] of data from two large multicenter clinical trials appears at odds with this conclusion; healthy premenopausal women assessed to be at high risk for breast cancer who had a high BMI were found to be at an increased risk for breast cancer. However, there are two points to note: all of the women being aged 35 years or older, the result is not consistent with observation by Peacock et al. [12], and consideration need to be given to the potential for a modifying influence of the designated markers of increased breast cancer risk in these prevention trials.

3.3.2 Postmenopause

There is general agreement that obesity is associated with an increased incidence of breast cancer in postmenopausal women (reviewed in [14–17]). The European Prospective Investigation into Cancer and Nutrition (EPIC) study [18], which had

57,923 postmenopausal participants, is of particular interest because of its large size, its prospective design, and the observations made concerning exogenous estrogens as a confounder. The results showed that a long-term weight gain was related to an increase in risk, but only in those who were not taking hormone replacement medication: compared with women with a stable body weight the relative risk for women who gained 15–20 kg was 1.5 with a confidence interval of 1.60–2.13. As reported by others, adiposity ceased to be a risk factor in current replacement therapy users, who were already at a high risk for breast cancer compared with nonusers.

Studies have been reported which failed to demonstrate a positive relationship between adiposity and postmenopausal breast cancer risk, but this is most likely an issue of experimental design; they were case–control studies [19–22], based on a single and often self-reported BMI determination, which had been made at or close to the time of diagnosis with the potential for modification due to the disease process and related therapy. Another cause for concern is failure to take account of the racial and ethnic composition of the study population; in the majority of the reported studies, most participants have been European or North American whites, but differences have been reported when compared with African-American women ([21, 23] and the references therein).

3.4 Obesity and Breast Cancer Prognosis

Preexisting obesity and postoperative weight gain are associated with poor prognosis in both premenopausal and postmenopausal breast cancer patients. In one study by Berclaz et al. [24] of 6,792 women who had participated in the International Breast Cancer Study Group investigation, those who were overweight (BMI: 25.0–29.9) or obese had significantly shorter overall and disease-free survivals than those with lower BMI values. A pivotal review of the literature by Chlebowski et al. [25] found that in 26 out of 34 studies individual studies, totaling 29,460 women, obesity was related to an increased risk of recurrence or reduced survival. Elsewhere [8], we have discussed obesity and the breast cancer cell estrogen receptor (ER) as a biomarker of disease prognosis in some depth (Table 3.1). Briefly, ER expression is predictive a good clinical outcome; it is more often present in breast cancer patients who are beyond the menopause and is associated with well-differentiated tumors with lower cell proliferation rates, themselves biomarkers of a good prognosis; but, paradoxically, ER-positive breast cancer is more prevalent in obese than nonobese women.

Estrogen-independent, ER-negative tumors predominate in obese premenopausal women [26–28]. Daling et al. [29] have provided a major contribution to our understanding in the relationships between body fat mass and tumor biomarkers of progression in young breast cancer patients. In their study, not only was a combination of obesity and an absence of ER expression in premenopausal breast cancer patients aged younger than 45 years associated with an increased risk of dying from the disease, but those with BMI values in the highest quartile were more likely to have larger tumors of high histologic grade. This observation is particularly

Table 3.1 Features of the adverse effects of obesity on prognosis in premenopausal and postmenopausal breast cancer patients

Premenopausal	Postmenopausal
Estrogen-independent, ER-negative tumors predominate	Estrogen-dependent, ER-positive tumors predominate
Association of obesity with poor prognosis reported to be stronger before menopause, but this remains uncertain	Antiestrogen therapy may negate the estrogen-associated adverse effect of obesity in ER-positive breast cancer
Triple-negative breast cancers more common and associated with obesity	Triple-negative breast cancer less common after menopause
Obesity associated with aggressive tumors with high cell proliferation rates	Higher BMI associated with lower proliferation rates
Mechanisms probably involve protein hormones, such as insulin and growth factors, such as adipokines, rather than estrogens	A major mechanistic factor is likely to be related to increased extraglandular estrogen production

ER estrogen receptor, *BMI* body mass index

significant because it implies that large tumors in overweight/obese women grow at a faster rate than tumors of similar size from leaner women, rather than simply arising from delayed diagnosis due to palpation difficulty in obese women.

The differences in breast cancer risk and prognosis with increasing age and adiposity most likely relate to changes in the estrogenic sex steroid hormones. As we discuss later (Sect. 7.7.1), in postmenopausal women, the proliferation of ER-positive breast cancer cells is stimulated by estrogens, which are synthesized in adipose stromal cells, increased production of which occurs in obesity. Before the menopause, the circulating estrogens are produced very largely by the ovaries, and one explanation hypothesized for the protective effect of obesity against premenopausal breast cancer is the known association with anovulatory menstrual cycles and reduced ovarian estrogen production. On this basis, the development of ER-negative breast cancer results from estrogen independence with growth promotion arising from the stimulatory activity of growth factors such as insulin [30], insulin-like growth factors, and some adipokines, notably leptin [31].

3.5 Upper Versus Lower Body Obesity and Breast Cancer Risk

The distribution of body fat, as opposed to general obesity, has also been related to the breast cancer risk. Typically, body fat locates to either the upper abdominal part of the body or to lower sites around the hips and thighs. The ratio of the waist-to-hip circumference (WHR) has been the most frequently used measurement to assess body fat distribution with upper body, or “central,” obesity being represented by a high ratio.

The general consensus is that the positive influence of adiposity on postmenopausal breast cancer risk and outcome applies specifically to upper body obesity [16, 17, 32–35]. Harvie et al. [16] reported a 34 % reduction in breast cancer risk for postmenopausal women with the lowest WHR values unadjusted for the BMI, but no influence of the WHR when considered alone in premenopausal women. However, when the effect of general obesity was eliminated by adjusting for the BMI, not only was the risk effect for the WHR diminished in postmenopausal women, but also in those who were still premenopausal there was now a 37 % reduction in risk for those with the lowest WHRs. The authors pointed out that the WHR on postmenopausal breast cancer risk is largely due to the coexistence of generalized obesity, whereas local fat accumulation with an upper body distribution (central obesity) specifically increases breast cancer risk in premenopausal women. Most studies using WHR have found that upper body obesity was a risk factor for postmenopausal breast cancer. Upper body obesity and an elevated WHR are also risk factors for another estrogen-related cancer, carcinoma of the endometrium, and T2D.

3.6 Type 2 Diabetes and Breast Cancer Risk and Prognosis

Wolf et al. [36] and Schott et al. [37] suggested that up to 16 % of breast cancer patients have diabetes, and that T2D may be associated with a 10–20 % excessive risk of breast cancer. Type 2 diabetes and breast cancer are two of the most common potentially life-threatening diseases among women in the developed and developing countries of the world. The effect that the present trend for T2D and obesity to occur with increasing frequency in children and young adults will have in relation to breast cancer is difficult to predict.

3.6.1 Obesity-Related Variables

There is ample epidemiological evidence that diabetes contributes to breast cancer risk [17, 36–40]. A summary and meta-analysis by Larsson and his colleagues [41] of 20 studies performed in 9 different countries revealed that women with obesity were more likely to be diagnosed with T2D; postmenopausal women with diabetes had a significant, 20 %, increase in breast cancer risk, but this was independent of the BMI [41]. There was no relationship between diabetes and premenopausal breast cancer in this study.

In Table 3.2, we have summarized a number of prospective and case–control studies that show breast cancer risk in patients after adjustment for nine factors, including BMI and menopausal status [42–45]. Overall, they are in agreement that T2D is a positive risk factor for postmenopausal breast cancer. The larger prospective United States Nurses' Health Study [42], which acquired 5,189 incident invasive breast cancer cases, showed a positive association with a history of T2D in

Table 3.2 Studies of breast cancer risk in patients with diabetes adjusted for body mass index (BMI) and/or menopausal status

Authors [reference]	Trial design	Adjusted relative risk (95 % confidence interval)	Adjustment BMI	Menopausal status
Michels et al. [42]	Prospective	1.17 (1.01–1.35)	Yes	Yes
Talamini et al. [43]	Case–control	1.4 (1.0–1.8)	Yes	Yes
Mink et al. [44]	Prospective	1.39 (0.86–2.23)	Yes	No
Franceschi et al. [45]	Case–control	1.0 (0.8–1.3)	Yes	Yes
Rollison et al. [46]	Case–control	1.06[a] (0.85–1.32)	Yes	Yes

[a]Odds ratio, after adjustment for nine factors, including BMI and menopausal status

postmenopausal women and without any modifying influence of the BMI. The only dissenting report in Table 3.2 is from a case–control study performed by Rollison et al. [46], which was designed to examine T2D as a risk factor for breast cancer in Hispanic compared with white American women; it showed no association in either ethnic group, although the Hispanic women, as expected from previous comparisons, were at a relatively high risk of T2D and lower breast cancer risk than non-Hispanic white women.

Overall survival in cancer patients, with or without preexisting diabetes, has shown diabetes to be associated with an increased all-cause mortality risk. Barone et al. identified 23 studies of diabetes and cancer prognosis that fulfilled their criteria of suitability for further evaluation by meta-analysis [47]. The same group examined the association between T2D in breast cancer patients and all-cause mortality risk [48]. They did show a 49 % higher risk of all-cause mortality in the patients with both conditions (hazard ratio, 1.49; 95 % CI, 1.35–1.65). One problem that occurs in the interpretation of data from studies that use large population databases, and all-cause mortality, is the coexistence of diabetes with a high prevalence of comorbidities unrelated to the breast cancer. The Danish Breast Cancer Cooperative Group, with 18,762 newly diagnosed T2D cases, found that the recurrence with metastases was 46 % higher in obese women with a BMI of 30 kg/m^2 or greater beyond the first 5 years. Obesity-related variables, such as body weight, BMI, and waist-and-hip circumference, have been associated with distant recurrence and death. This has been confirmed by Goodwin et al. [49] who studied the outcomes of two cohort trials of women with localized, invasive breast cancer. This would imply that obesity plays a long-term, determinate role in the outcome of breast cancer.

3.6.2 *Insulin-Related Variables*

In previous prospective studies of c-peptide or insulin levels, elevated plasma levels have been related to disease recurrence and poor prognosis. The presence of T2D is consistent with hyperinsulinemia being a key biological mechanism by which the diabetic state promotes breast cancer development and subsequent progression [50]. However, before the menopause, relatively high circulating levels of insulin may be associated with reduced breast cancer risk is consistent with the apparently protective effect of diabetes observed by Baron et al. [47] and is similar to what is observed with obesity. However, Goodwin et al. [51] found that elevated plasma insulin concentrations in both nondiabetic premenopausal and postmenopausal breast cancer patients were associated with an increased risk of recurrence at distant sites and with a poor prognosis: a positive correlation between the insulin levels and BMI. Interestingly, the relationship between the hormone and disease outcome persisted even after adjustment for adiposity. A similar relationship has been observed between breast cancer prognosis and fasting serum C-peptide levels in women with ER-positive tumors [52].

3.7 Prostate Cancer

Prostate cancer may be a low-grade, relatively indolent, neoplasm that develops slowly and has low invasive potential, and, indeed may never progress to become a clinically significant disease. Alternatively, this tumor type may be biologically aggressive, with high histologic grade and a propensity for early invasion beyond the prostatic capsule and metastasis to distant sites. Whether the two forms always share the same risk factors is unknown, but, as we shall see, for obesity this does not appear to be the case.

3.7.1 Obesity

The relationship between obesity and prostate cancer is a complicated one. Some earlier investigations of prostate cancer risk showed a positive association [53], but more recently, others, including two prospective studies from Sweden [54] and the USA [55], with 9,000 and 47,781 participants, respectively, found no association. The explanation for this confusion may rest, at least in part, in the reports that obesity as a positive risk factor for prostate cancer relates specifically with the aggressive phenotype [56–60], an association with significant implications for prostate cancer pathology and therapy. In support, a meta-analysis by Discacciati et al. [61] of the results from 25 studies that examined disease stage and the BMI showed not only a positive relationship between obesity and advanced prostate cancer but also a decrease in the risk for localized disease.

The association between obesity and an aggressive prostate cancer phenotype is reflected in the relationship between the BMI and prostate cancer mortality rate. For example, in one large retrospective cohort study by Andersson et al. [62], which showed only a trend for an association between the BMI and overall prostate cancer incidence (P for trend = 0.01), there was a significantly larger prostate cancer mortality rate in the higher BMI categories (P for trend = 0.040). Similarly, a prospective study of 287,760 men, aged 50–71 years at entry and with a 6-year follow-up, showed a significant positive relationship between mortality risk and the baseline BMI (*P* for trend = 0.02), whereas higher BMI values were associated with a *reduced* prostate cancer incidence that the investigators ascribed to the influence of indolent localized disease [63]. Amling et al. [56] found that obese patients with aggressive prostate cancer presented at a relatively young age, and that African-American prostate cancer patients had higher BMI values compared with other racial and ethnic groups and were more likely to have high-grade tumors. This association of obesity with a young age at diagnosis and high-grade prostate cancers has also been observed by other investigators [64].

The obesity associated with an increased risk of aggressive cancer of the prostate has the same body composition as that related to breast cancer. A prospective study from Italy [59], which found an association between obesity and high-grade prostate cancer, also showed that the relationship was with upper body adiposity as judged by an elevated waist circumference. A second prospective study [65] from Sweden demonstrated a similar relationship when the WHR was used to assess fat distribution, which was independent of BMI.

3.7.2 Diabetes

Two studies have been reported in which meta-analysis was used to examine previously published investigations into the relationship between diabetes mellitus and prostate cancer risk [66, 67]. The available data did not permit a distinction between type 1 and 2 diabetes, but the reasonable assumption was made that the analysis represented T2D, given the age range of the populations and the fact that 90–95 % of diabetics have this form of the disease. Bonovas et al. [66] performed their literature search up to the year 2003 and identified 14 studies, 9 cohort and 5 case–control in design, that met their criteria for inclusion. The meta-analysis showed that there was an *inverse* relationship between diabetes and prostate cancer risk, which translated to a 9 % reduction in risk. One problem with the analysis is that no examination was made of obesity as a potential confounder, although it has been calculated that as many as 90.4 % of cases, T2D is associated with the presence of a high BMI [68].

The meta-analysis performed by Kaspar and Giovannucci [67] was published 2 years after that of Bonovas and his coworkers. There were 19 studies, which provided more than twice the number of prostate cancer cases; 12 had also been included in the analysis of Bonovas et al. [66]. The overall conclusion was the same: diabetic men have a significantly *decreased* risk of developing prostate cancer (RR = 0.84; 95 % CI, 0.76–0.93). A particularly important contribution made by Kaspar and Giovannucci was that they did examine the potential interaction with obesity. In 12 of the 19 studies, there was no adjustment for the BMI and the calculated RR was 0.87 (95 % CI, 0.76–0.99); when an adjustment was made, it had no significant effect on the observed relationship between a history of diabetes and prostate cancer risk: RR, 0.82 (95 % CI, 0.69–0.97).

Gong et al. [69] reported a large prospective study of diabetes and prostate cancer from the USA after the two meta-analyses described above had been published that also took account of potential confounding by obesity. Men with diabetes had a 34 % lower risk of prostate cancer compared with men without diabetes that was not affected by adjustment for the BMI; the association was stronger for low-grade than for high-grade tumors. Zhu et al. [70] used a nested case–control design with 1,110 prostate cancer patients diagnosed in the period 1982–1995. Logistic regression analysis showed that a history of diabetes was related to a decreased risk of prostate; the BMI did not differ significantly in the prostate cancer compared with the control group. In contrast to these results, recently published studies have found that the presence of diabetes is *positively* associated with prostate cancers of high-grade [71–73] and late-stage tumors [72], a reversal in the observed relationship that needs to be considered in the context of the duration of the presence of T2D and the detection of prostate cancer by prostatic-specific antigen screening.

3.8 Mechanisms

Generalized upper body (central or visceral) adiposity influences the development of breast and prostate cancers by mechanisms, which are of endocrine origin, while locally the breast and periprostatic adipose tissues do so as the principal sources of hormones and adipokines concerned with paracrine mechanisms of action.

3.8.1 Sex Steroid Hormones

The estrogens involved in the stimulation of breast cancer development are produced before the menopause by the ovaries, but in postmenopausal women they are synthesized almost exclusively in adipose tissue stromal cells by the enzymatic aromatization of the C19 steroid androstenedione (Fig. 3.1). Obese postmenopausal women have elevated levels of estrogens and these can promote ER-positive breast cancer development and stimulation of metastasis by both endocrine and paracrine mechanisms [74]. An endocrine role is supported by the demonstration that an elevation in plasma estrogen concentrations is a risk factor for postmenopausal breast cancer [23, 75], but, in addition, particularly high levels of aromatase activity are often present in the adipose tissue fibroblasts that form a dense capsule around a premalignant or cancerous breast lesion and this source of estrogen can create a local microenvironment, which favors growth stimulation by a paracrine mechanism [76].

The biological activity of the sex steroid hormones is closely regulated and under physiological circumstances 30–50 % of the plasma estradiol is tightly bound to sex hormone-binding globulin (SHBG), which renders it biologically inert. Low plasma SHBG concentrations occur in obesity and result in elevated levels of

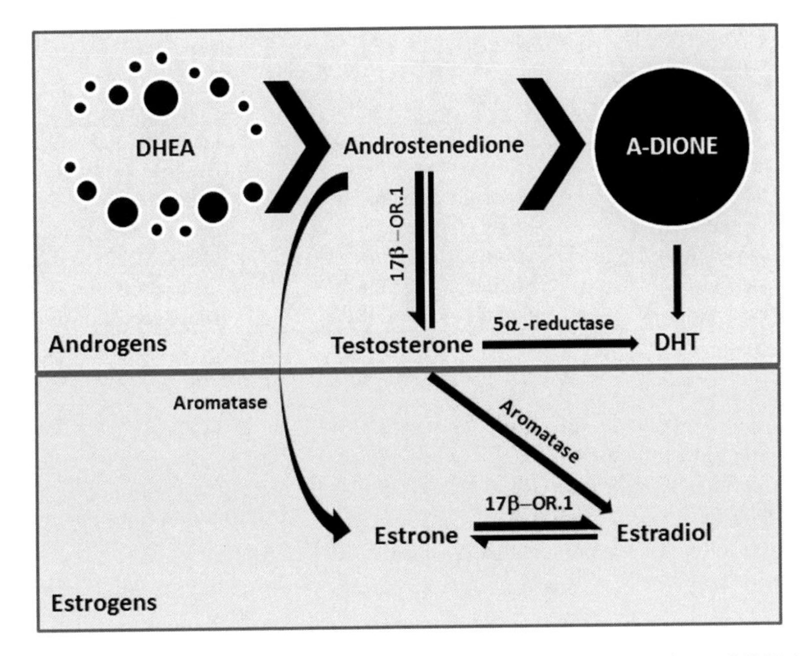

Fig. 3.1 Androgen metabolism and estrogen synthesis in abdominal adipose tissue. *DHEA* dehydroepiandrosterone, *A-DIONE* 5α-androstane-3,17-dione, *DHT* dihydrotestosterone, *17βOR.1* 17β-oxidoreductase type 1

bioactive estradiol changes, which have also been associated with increased breast cancer risk [23, 75].

Testosterone binds to SHBG with an even greater affinity than estradiol and, again, reduced levels of the binding-globulin result in elevations in the biologically available androgen. High plasma concentrations of the total [77, 78] and nonprotein bound, biologically available [77], testosterone have been associated with an increase in postmenopausal breast cancer risk. Kaaks et al. [23] also determined the serum concentrations of androstenedione and dehydroepiandrosterone (DHEA), which in postmenopausal women are not extraglandular in origin but are secreted by the adrenal glands (Fig. 3.1) and found that the concentrations of both were positively correlated with breast cancer risk. However, unlike the estrogens and nonprotein-bound testosterone, there was no association with the BMI.

Type 2 diabetes in postmenopausal women is frequently accompanied by an elevation in the plasma estrogens, which is attributable only in part to obesity [79, 80]. Kalyani et al. [80] found that there was a positive association between the serum estradiol concentration and T2D risk, and a negative one for the SHBG, both of which were weakened, but not reversed, by adjustment for the BMI. The extent to which these changes in estrogen activity contribute to the increased risk of breast cancer in postmenopausal diabetic women is unknown, and of more biological significance may be the high prevalence of obesity in T2D.

Table 3.3 Comparison of leptin and adiponectin: metabolic associations and effects on breast and prostate cancer

	Leptin	Adiponectin
Obesity	Plasma levels elevated	Plasma levels decreased
Insulin function	Decreased sensitivity	Decreased resistance
Type 2 diabetes	Promotes development	Reduces risk
Breast cancer risk	Plasma levels: equivocal	Plasma levels: reduced risk
Prostate cancer risk	Plasma levels: equivocal	Plasma levels: reduced risk
Breast cancer growth	Stimulates cell growth	Inhibits cell growth
Prostate cancer growth	Stimulates cell growth	Inhibits cell growth

Platz and Giovannucci [81] found a lack of consistency in the results obtained in published studies of associations between prostate cancer risk and blood testosterone and dihydrotestosterone (DHT) concentrations. They pointed out that the circulating androgen concentrations probably do not reflect the situation within the periprostatic fat and the prostate itself for either rate of conversion of testosterone to DHT, the biologically potent androgenic steroid, or the tissue concentrations and level of androgen action on the prostatic epithelial cells. Also, there were the usual concerns, which apply equally to the estrogens, regarding the reliability of studies using single serum samples for hormone assays, and, in prospective studies, the appropriate time in life for making an assessment of endocrine status and future prostate cancer risk. Despite these limitations, Gann et al. [82], in a large prospective, nested case–control study of American physicians, found that when the plasma testosterone and SHBG were considered in combination, so as to negate any mutual confounding effect, increasing levels of testosterone and decreasing levels of SHBG showed strong trends for increasing prostate cancer risk. The higher levels of testosterone obtained were still within the normal range, but they were associated with an approximately 2.5-fold increase in the risk of developing prostate cancer after adjustment for the plasma SHBG and estradiol. A nonlinear inverse association of prostate cancer risk with increasing estradiol levels was also found in this study.

3.8.2 Adipokines

The adipokines are a group of proteins synthesized in the adipocytes and stromal cells of adipose tissue and in the macrophages that infiltrate the adipose cell mass. Leptin and adiponectin are multifunctional adipokines, which have been studied extensively in breast and, to a lesser degree, prostate cancer, and which have very different, largely opposing, physiological and pathological functions (Table 3.3). They both circulate in the blood and so can potentially exert their biological effects in an endocrine manner; in addition, their local production in mammary and periprostatic fat provides for paracrine action on adjacent target cells that possess the necessary receptors.

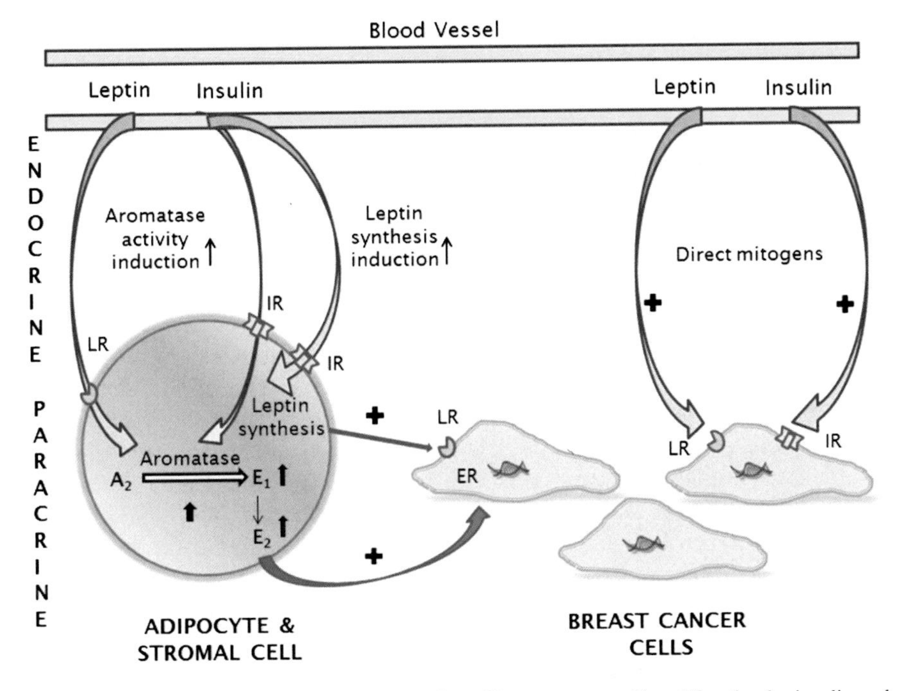

Fig. 3.2 The endocrine and paracrine stimulation of breast cancer cell proliferation by insulin and leptin. High plasma insulin and leptin levels, a consequence of generalized upper body obesity, can act directly in an endocrine manner as breast cancer cell mitogens; they also induce adipose stromal cell aromatase expression with resulting increased estrogen production and paracrine stimulation of breast cancer proliferation. Insulin also stimulates the production of leptin by adipocytes, so increasing further the local availability of the adipokine for paracrine activity. *A2* androstenedione, *E1* estrone, *E2* estradiol, *ER* estrogen receptor, *LR* leptin receptor

The plasma leptin concentrations correlate positively with the BMI and are high in obese women, but studies designed to demonstrate an association with breast cancer risk has produced conflicting results (reviewed in [74]). Furthermore, when Goodwin et al. [83] evaluated plasma leptin as a prognostic biomarker in breast cancer patients, they found that there was no association with disease outcome, although levels were higher in patients with high-grade and ER-negative tumors. Paracrine loops between cells of the breast adipose tissue and breast epithelial cells have a major role in the relationship between adiposity and breast cancer development and progression (Fig. 3.2), and this appears to include the mode of action of leptin [74]. Thus, the expression of leptin protein has been demonstrated in adipocytes that are in close contact with breast cancer cells [84], a tumor cell type that expresses leptin receptors, and for which leptin is a mitogen [85]. Circulating leptin levels are elevated in insulin-resistant patients with T2D [76]. This is independent of the BMI and may be a reflection of the stimulation of leptin production, which occurs as a result of the accompanying hyperinsulinemia [86]; insulin has been shown to stimulate leptin secretion from adipocytes [87] and so may be involved in the paracrine mechanism by which the adipokine promotes breast cancer cell growth.

Plasma adiponectin concentrations are *reduced* in obese women, and, in support of an endocrine role, epidemiological studies have unequivocally demonstrated an inverse relationship with postmenopausal breast cancer risk (reviewed in [74]). Tworoger et al. [88] confirmed the relationship with postmenopausal status and also showed that it was stronger in women who had never used hormone replacement therapy and those with low plasma estrogen concentrations. Plasma adiponectin levels are also reduced in type 2 diabetes, a relationship that is independent of the BMI; in one study, the extent of the reduction in the plasma adiponectin corresponded more closely to the severity of insulin resistance and the elevation in serum insulin than to the degree of adiposity [89].

Most studies of prostate cancer found no evidence of an association between the plasma leptin concentration and either disease risk [53, 90, 91] or stage at diagnosis [92], but a reverse relationship has been reported for the plasma adiponectin [77, 91], and particularly so in the case of high-grade tumors [23]. The periprostatic adipose tissue provides a site for adipokine production and a local source of leptin for the establishment of a paracrine loop. Prostate cancer cells do possess the receptors required for such a mechanism, and leptin does stimulate the growth and migration of some human prostate cancer cell lines in vitro [53, 93]. Moreover, a positive relationship has been demonstrated between the amount of periprostatic fat in prostate cancer patients and the risk of aggressive disease [94].

3.8.3 *Insulin*

Insulin produced in the pancreatic β cells can exert stimulatory effects on breast cancer epithelial cells by endocrine, paracrine, and autocrine mechanisms (Fig. 3.2). It is a potent direct mitogen for breast cancer cells, may contribute to tumor growth by inducing the aromatase that catalyzes estrone production from androstenedione and enhancing the bioavailability of estradiol (reviewed in [95]), and is also a vascular endothelial cell growth factor and so may stimulate tumor-related angiogenesis.

The circumstances under which T2D results in either a decrease or increase in prostate carcinogenesis may be related to endocrine changes over time; hyperinsulinemia is present in the "prediabetic" and early clinical stages of the disease, but later a loss of β-cell function occurs and serum insulin concentrations fall to subnormal levels. High serum insulin concentrations have been associated with increased prostate cancer risk, and insulin is a mitogen for prostate cancer cells in vitro [96]. With these observations in mind, Kasper et al. [97] used data from a prospective cohort to examine the changes in prostate cancer risk with time since a diagnosis of diabetes had been made. They found that there was no reduction in prostate cancer risk in the first year [hazard ratio was 0.78 (CI, 0.63–0.96; P for trend = 0.01)]. It is to be hoped that in the future the opportunity will arise for a confirmatory study that includes serial serum insulin assays.

3.9 Concluding Remarks

The expansion of white adipose tissue directly contributes to the pathogenesis of obesity and is linked to the progression of breast and prostate cancers. Adiposity, together with type 2 diabetes, promotes an aggressive and metastatic phenotype via endocrine and paracrine mechanisms of action by increasing the expression patterns of hormones and adipokines, both locally and systemically, and altering their signaling networks to support the tumor microenvironment.

References

1. American Cancer Society. Cancer facts and figures 2010. Atlanta: American Cancer Society. 12 Dec 2011 [cited 12 Dec 2011]. Available from http://cancer.org
2. Hulka BS, Liu ET, Lininger RA. Steroid hormones and risk of breast cancer. Cancer. 1994;74(3 Suppl):1111–24.
3. Nacusi LP, Tindall DJ. Targeting 5alpha-reductase for prostate cancer prevention and treatment. Nat Rev Urol. 2011;8(7):378–84.
4. Bentzon N, During M, Rasmussen BB, Mouridsen H, Kroman N. Prognostic effect of estrogen receptor status across age in primary breast cancer. Int J Cancer. 2008;122(5):1089–94.
5. Massard C, Fizazi K. Targeting continued androgen receptor signaling in prostate cancer. Clin Cancer Res. 2011;17(12):3876–83.
6. Flegal KM, Carroll MD, Ogden CL, Curtin LR. Prevalence and trends in obesity among US adults, 1999–2008. JAMA. 2010;303(3):235–41.
7. Cowie CC, Rust KF, Byrd-Holt DD, Eberhardt MS, Flegal KM, Engelgau MM, et al. Prevalence of diabetes and impaired fasting glucose in adults in the U.S. population: National Health and Nutrition Examination Survey 1999–2002. Diabetes Care. 2006;29(6):1263–8.
8. Rose DP, Vona-Davis L. Influence of obesity on breast cancer receptor status and prognosis. Expert Rev Anticancer Ther. 2009;9(8):1091–101.
9. Vona-Davis L, Rose DP. Type 2 diabetes and obesity metabolic interactions: common factors for breast cancer risk and novel approaches to prevention and therapy. Curr Diabetes Rev. 2012;8(2):116–30.
10. Ursin G, Longnecker MP, Haile RW, Greenland S. A meta-analysis of body mass index and risk of premenopausal breast cancer. Epidemiology. 1995;6(2):137–41.
11. van den Brandt PA, Spiegelman D, Yaun SS, Adami HO, Beeson L, Folsom AR, et al. Pooled analysis of prospective cohort studies on height, weight, and breast cancer risk. Am J Epidemiol. 2000;152(6):514–27.
12. Peacock SL, White E, Daling JR, Voigt LF, Malone KE. Relation between obesity and breast cancer in young women. Am J Epidemiol. 1999;149(4):339–46.
13. Cecchini RS, Costantino JP, Cauley JA, Cronin WM, Wickerham DL, Land SR, et al. Body mass index and the risk for developing invasive breast cancer among high-risk women in NSABP P-1 and STAR breast cancer prevention trials. Cancer Prev Res (Phila). 2012;5(4): 583–92.
14. McTiernan A. Associations between energy balance and body mass index and risk of breast carcinoma in women from diverse racial and ethnic backgrounds in the U.S. Cancer. 2000;88(5 Suppl):1248–55.
15. Stephenson GD, Rose DP. Breast cancer and obesity: an update. Nutr Cancer. 2003;45(1): 1–16.
16. Harvie M, Hooper L, Howell AH. Central obesity and breast cancer risk: a systematic review. Obes Rev. 2003;4(3):157–73.

17. Lorincz AM, Sukumar S. Molecular links between obesity and breast cancer. Endocr Relat Cancer. 2006;13(2):279–92.
18. Lahmann PH, Hoffmann K, Allen N, van Gils CH, Khaw KT, Tehard B, et al. Body size and breast cancer risk: findings from the European Prospective Investigation into Cancer and Nutrition (EPIC). Int J Cancer. 2004;111(5):762–71.
19. John EM, Sangaramoorthy M, Phipps AI, Koo J, Horn-Ross PL. Adult body size, hormone receptor status, and premenopausal breast cancer risk in a multiethnic population: the San Francisco Bay Area breast cancer study. Am J Epidemiol. 2011;173(2):201–16.
20. Berstad P, Coates RJ, Bernstein L, Folger SG, Malone KE, Marchbanks PA, et al. A case-control study of body mass index and breast cancer risk in white and African-American women. Cancer Epidemiol Biomarkers Prev. 2010;19(6):1532–44.
21. Ogundiran TO, Huo D, Adenipekun A, Campbell O, Oyesegun R, Akang E, et al. Case-control study of body size and breast cancer risk in Nigerian women. Am J Epidemiol. 2010;172(6): 682–90.
22. Boyd NF, Martin LJ, Sun L, Guo H, Chiarelli A, Hislop G, et al. Body size, mammographic density, and breast cancer risk. Cancer Epidemiol Biomarkers Prev. 2006;15(11):2086–92.
23. Kaaks R, Rinaldi S, Key TJ, Berrino F, Peeters PH, Biessy C, et al. Postmenopausal serum androgens, oestrogens and breast cancer risk: the European prospective investigation into cancer and nutrition. Endocr Relat Cancer. 2005;12(4):1071–82.
24. Berclaz G, Li S, Price KN, Coates AS, Castiglione-Gertsch M, Rudenstam CM, et al. Body mass index as a prognostic feature in operable breast cancer: the International Breast Cancer Study Group experience. Ann Oncol. 2004;15(6):875–84.
25. Chlebowski RT, Chen Z, Anderson GL, Rohan T, Aragaki A, Lane D, et al. Ethnicity and breast cancer: factors influencing differences in incidence and outcome. J Natl Cancer Inst. 2005;97(6):439–48.
26. Colditz GA, Rosner BA, Chen WY, Holmes MD, Hankinson SE. Risk factors for breast cancer according to estrogen and progesterone receptor status. J Natl Cancer Inst. 2004;96(3): 218–28.
27. Suzuki R, Orsini N, Saji S, Key TJ, Wolk A. Body weight and incidence of breast cancer defined by estrogen and progesterone receptor status–a meta-analysis. Int J Cancer. 2009; 124(3):698–712.
28. Kyndi M, Sorensen FB, Knudsen H, Overgaard M, Nielsen HM, Overgaard J. Estrogen receptor, progesterone receptor, HER-2, and response to postmastectomy radiotherapy in high-risk breast cancer: the Danish Breast Cancer Cooperative Group. J Clin Oncol. 2008;26(9): 1419–26.
29. Daling JR, Malone KE, Doody DR, Johnson LG, Gralow JR, Porter PL. Relation of body mass index to tumor markers and survival among young women with invasive ductal breast carcinoma. Cancer. 2001;92(4):720–9.
30. Kerbel RS. Tumor angiogenesis. N Engl J Med. 2008;358(19):2039–49.
31. Liu CL, Chang YC, Cheng SP, Chern SR, Yang TL, Lee JJ, et al. The roles of serum leptin concentration and polymorphism in leptin receptor gene at codon 109 in breast cancer. Oncology. 2007;72(1–2):75–81.
32. Friedenreich CM. Review of anthropometric factors and breast cancer risk. Eur J Cancer Prev. 2001;10(1):15–32.
33. Connolly BS, Barnett C, Vogt KN, Li T, Stone J, Boyd NF. A meta-analysis of published literature on waist-to-hip ratio and risk of breast cancer. Nutr Cancer. 2002;44(2):127–38.
34. Onat A, Avci GS, Barlan MM, Uyarel H, Uzunlar B, Sansoy V. Measures of abdominal obesity assessed for visceral adiposity and relation to coronary risk. Int J Obes Relat Metab Disord. 2004;28(8):1018–25.
35. Carr DB, Utzschneider KM, Hull RL, Kodama K, Retzlaff BM, Brunzell JD, et al. Intra-abdominal fat is a major determinant of the National Cholesterol Education Program Adult Treatment Panel III criteria for the metabolic syndrome. Diabetes. 2004;53(8):2087–94.
36. Wolf I, Sadetzki S, Catane R, Karasik A, Kaufman B. Diabetes mellitus and breast cancer. Lancet Oncol. 2005;6(2):103–11.

37. Schott S, Schneeweiss A, Sohn C. Breast cancer and diabetes mellitus. Exp Clin Endocrinol Diabetes. 2010;118(10):673–7.
38. Xue F, Michels KB. Diabetes, metabolic syndrome, and breast cancer: a review of the current evidence. Am J Clin Nutr. 2007;86(3):s823–35.
39. Rose DP, Haffner SM, Baillargeon J. Adiposity, the metabolic syndrome, and breast cancer in African-American and white American women. Endocr Rev. 2007;28(7):763–77.
40. Vigneri P, Frasca F, Sciacca L, Pandini G, Vigneri R. Diabetes and cancer. Endocr Relat Cancer. 2009;16(4):1103–23.
41. Larsson SC, Mantzoros CS, Wolk A. Diabetes mellitus and risk of breast cancer: a meta-analysis. Int J Cancer. 2007;121(4):856–62.
42. Michels KB, Solomon CG, Hu FB, Rosner BA, Hankinson SE, Colditz GA, et al. Type 2 diabetes and subsequent incidence of breast cancer in the Nurses' Health Study. Diabetes Care. 2003;26(6):1752–8.
43. Talamini R, Franceschi S, Favero A, Negri E, Parazzini F, La Vecchia C. Selected medical conditions and risk of breast cancer. Br J Cancer. 1997;75(11):1699–703.
44. Mink PJ, Shahar E, Rosamond WD, Alberg AJ, Folsom AR. Serum insulin and glucose levels and breast cancer incidence: the atherosclerosis risk in communities study. Am J Epidemiol. 2002;156(4):349–52.
45. Franceschi S, La VC, Negri E, Parazzini F, Boyle P. Breast cancer risk and history of selected medical conditions linked with female hormones. Eur J Cancer. 1990;26(7):781–5.
46. Rollison DE, Giuliano AR, Sellers TA, Laronga C, Sweeney C, Risendal B, et al. Population-based case-control study of diabetes and breast cancer risk in Hispanic and non-Hispanic White women living in US southwestern states. Am J Epidemiol. 2008;167(4):447–56.
47. Barone BB, Yeh HC, Snyder CF, Peairs KS, Stein KB, Derr RL, et al. Long-term all-cause mortality in cancer patients with preexisting diabetes mellitus: a systematic review and meta-analysis. JAMA. 2008;300(23):2754–64.
48. Peairs KS, Barone BB, Snyder CF, Yeh HC, Stein KB, Derr RL, et al. Diabetes mellitus and breast cancer outcomes: a systematic review and meta-analysis. J Clin Oncol. 2011;29(1):40–6.
49. Goodwin PJ, Ennis M, Pritchard KI, Trudeau ME, Koo J, Taylor SK, et al. Insulin- and obesity-related variables in early-stage breast cancer: correlations and time course of prognostic associations. J Clin Oncol. 2012;30(2):164–71.
50. Vona-Davis L, Howard-McNatt M, Rose DP. Adiposity, type 2 diabetes and the metabolic syndrome in breast cancer. Obes Rev. 2007;8(5):395–408.
51. Goodwin PJ, Ennis M, Pritchard KI, Trudeau ME, Koo J, Madarnas Y, et al. Fasting insulin and outcome in early-stage breast cancer: results of a prospective cohort study. J Clin Oncol. 2002;20(1):42–51.
52. Irwin ML, Duggan C, Wang CY, Smith AW, McTiernan A, Baumgartner RN, et al. Fasting C-peptide levels and death resulting from all causes and breast cancer: the health, eating, activity, and lifestyle study. J Clin Oncol. 2011;29(1):47–53.
53. Baillargeon J, Rose DP. Obesity, adipokines, and prostate cancer (review). Int J Oncol. 2006;28(3):737–45.
54. Jonsson F, Wolk A, Pedersen NL, Lichtenstein P, Terry P, Ahlbom A, et al. Obesity and hormone-dependent tumors: cohort and co-twin control studies based on the Swedish Twin Registry. Int J Cancer. 2003;106(4):594–9.
55. Giovannucci E, Rimm EB, Stampfer MJ, Colditz GA, Willett WC. Height, body weight, and risk of prostate cancer. Cancer Epidemiol Biomarkers Prev. 1997;6(8):557–63.
56. Amling CL, Kane CJ, Riffenburgh RH, Ward JF, Roberts JL, Lance RS, et al. Relationship between obesity and race in predicting adverse pathologic variables in patients undergoing radical prostatectomy. Urology. 2001;58(5):723–8.
57. Su LJ, Arab L, Steck SE, Fontham ET, Schroeder JC, Bensen JT, et al. Obesity and prostate cancer aggressiveness among African and Caucasian Americans in a population-based study. Cancer Epidemiol Biomarkers Prev. 2011;20(5):844–53.
58. Bassett JK, Severi G, Baglietto L, Macinnis RJ, Hoang HN, Hopper JL, et al. Weight change and prostate cancer incidence and mortality. Int J Cancer. 2012;131(7):1711–9.

59. De Nunzio C, Albisinni S, Freedland SJ, Miano L, Cindolo L, Finazzi AE, et al. Abdominal obesity as risk factor for prostate cancer diagnosis and high grade disease: a prospective multicenter Italian cohort study. Urol Oncol. 2011. http://dx.doi.org/10.1016/j.urolonc.2011.08.007.
60. De Nunzio C, Freedland SJ, Miano L, Finazzi AE, Banez L, Tubaro A. The uncertain relationship between obesity and prostate cancer: an Italian biopsy cohort analysis. Eur J Surg Oncol. 2011;37(12):1025–9.
61. Discacciati A, Orsini N, Wolk A. Body mass index and incidence of localized and advanced prostate cancer—a dose-response meta-analysis of prospective studies. Ann Oncol. 2012;23(7): 1665–71.
62. Andersson SO, Wolk A, Bergstrom R, Adami HO, Engholm G, Englund A, et al. Body size and prostate cancer: a 20-year follow-up study among 135006 Swedish construction workers. J Natl Cancer Inst. 1997;89(5):385–9.
63. Wright ME, Chang SC, Schatzkin A, Albanes D, Kipnis V, Mouw T, et al. Prospective study of adiposity and weight change in relation to prostate cancer incidence and mortality. Cancer. 2007;109(4):675–84.
64. Rohrmann S, Roberts WW, Walsh PC, Platz EA. Family history of prostate cancer and obesity in relation to high-grade disease and extraprostatic extension in young men with prostate cancer. Prostate. 2003;55(2):140–6.
65. Wallstrom P, Bjartell A, Gullberg B, Olsson H, Wirfalt E. A prospective Swedish study on body size, body composition, diabetes, and prostate cancer risk. Br J Cancer. 2009;100(11): 1799–805.
66. Bonovas S, Filioussi K, Tsantes A. Diabetes mellitus and risk of prostate cancer: a meta-analysis. Diabetologia. 2004;47(6):1071–8.
67. Kasper JS, Giovannucci E. A meta-analysis of diabetes mellitus and the risk of prostate cancer. Cancer Epidemiol Biomarkers Prev. 2006;15(11):2056–62.
68. Colditz GA, Willett WC, Stampfer MJ, Manson JE, Hennekens CH, Arky RA, et al. Weight as a risk factor for clinical diabetes in women. Am J Epidemiol. 1990;132(3):501–13.
69. Gong Z, Neuhouser ML, Goodman PJ, Albanes D, Chi C, Hsing AW, et al. Obesity, diabetes, and risk of prostate cancer: results from the prostate cancer prevention trial. Cancer Epidemiol Biomarkers Prev. 2006;15(10):1977–83.
70. Zhu K, Lee IM, Sesso HD, Buring JE, Levine RS, Gaziano JM. History of diabetes mellitus and risk of prostate cancer in physicians. Am J Epidemiol. 2004;159(10):978–82.
71. Abdollah F, Briganti A, Suardi N, Gallina A, Capitanio U, Salonia A, et al. Does diabetes mellitus increase the risk of high-grade prostate cancer in patients undergoing radical prostatectomy? Prostate Cancer Prostatic Dis. 2011;14(1):74–8.
72. Moreira DM, Anderson T, Gerber L, Thomas JA, Banez LL, McKeever MG, et al. The association of diabetes mellitus and high-grade prostate cancer in a multiethnic biopsy series. Cancer Causes Control. 2011;22(7):977–83.
73. Hong SK, Oh JJ, Byun SS, Hwang SI, Lee HJ, Choe G, et al. Impact of diabetes mellitus on the detection of prostate cancer via contemporary multi (>/= 12)-core prostate biopsy. Prostate. 2012;72(1):51–7.
74. Vona-Davis L, Rose DP. Adipokines as endocrine, paracrine, and autocrine factors in breast cancer risk and progression. Endocr Relat Cancer. 2007;14(2):189–206.
75. Zeleniuch-Jacquotte A, Shore RE, Koenig KL, Akhmedkhanov A, Afanasyeva Y, Kato I, et al. Postmenopausal levels of oestrogen, androgen, and SHBG and breast cancer: long-term results of a prospective study. Br J Cancer. 2004;90(1):153–9.
76. Fischer S, Hanefeld M, Haffner SM, Fusch C, Schwanebeck U, Kohler C, et al. Insulin-resistant patients with type 2 diabetes mellitus have higher serum leptin levels independently of body fat mass. Acta Diabetol. 2002;39(3):105–10.
77. Goktas S, Yilmaz MI, Caglar K, Sonmez A, Kilic S, Bedir S. Prostate cancer and adiponectin. Urology. 2005;65(6):1168–72.
78. Key T, Appleby P, Barnes I, Reeves G, Endogenous Hormones and Breast Cancer Collaborative Group. Endogenous sex hormones and breast cancer in postmenopausal women: reanalysis of nine prospective studies. J Natl Cancer Inst. 2002;94(8):606–16.

79. Ding EL, Song Y, Manson JE, Rifai N, Buring JE, Liu S. Plasma sex steroid hormones and risk of developing type 2 diabetes in women: a prospective study. Diabetologia. 2007;50(10):2076–84.
80. Kalyani RR, Franco M, Dobs AS, Ouyang P, Vaidya D, Bertoni A, et al. The association of endogenous sex hormones, adiposity, and insulin resistance with incident diabetes in postmenopausal women. J Clin Endocrinol Metab. 2009;94(11):4127–35.
81. Platz EA, Giovannucci E. The epidemiology of sex steroid hormones and their signaling and metabolic pathways in the etiology of prostate cancer. J Steroid Biochem Mol Biol. 2004;92(4):237–53.
82. Gann PH, Hennekens CH, Ma J, Longcope C, Stampfer MJ. Prospective study of sex hormone levels and risk of prostate cancer. J Natl Cancer Inst. 1996;88(16):1118–26.
83. Goodwin PJ, Ennis M, Fantus IG, Pritchard KI, Trudeau ME, Koo J, et al. Is leptin a mediator of adverse prognostic effects of obesity in breast cancer? J Clin Oncol. 2005;23(25):6037–42.
84. Celis JE, Moreira JM, Cabezon T, Gromov P, Friis E, Rank F, et al. Identification of extracellular and intracellular signaling components of the mammary adipose tissue and its interstitial fluid in high risk breast cancer patients: toward dissecting the molecular circuitry of epithelial-adipocyte stromal cell interactions. Mol Cell Proteomics. 2005;4(4):492–522.
85. Frankenberry KA, Skinner H, Somasundar P, McFadden DW, Vona-Davis LC. Leptin receptor expression and cell signaling in breast cancer. Int J Oncol. 2006;28(4):985–93.
86. Utriainen T, Malmstrom R, Makimattila S, Yki-Jarvinen H. Supraphysiological hyperinsulinemia increases plasma leptin concentrations after 4 h in normal subjects. Diabetes. 1996;45(10):1364–6.
87. Zeigerer A, Rodeheffer MS, McGraw TE, Friedman JM. Insulin regulates leptin secretion from 3T3-L1 adipocytes by a PI 3 kinase independent mechanism. Exp Cell Res. 2008;314(11–12):2249–56.
88. Tworoger SS, Eliassen AH, Kelesidis T, Colditz GA, Willett WC, Mantzoros CS, et al. Plasma adiponectin concentrations and risk of incident breast cancer. J Clin Endocrinol Metab. 2007;92(4):1510–6.
89. Weyer C, Funahashi T, Tanaka S, Hotta K, Matsuzawa Y, Pratley RE, et al. Hypoadiponectinemia in obesity and type 2 diabetes: close association with insulin resistance and hyperinsulinemia. J Clin Endocrinol Metab. 2001;86(5):1930–5.
90. Baillargeon J, Platz EA, Rose DP, Pollock BH, Ankerst DP, Haffner S, et al. Obesity, adipokines, and prostate cancer in a prospective population-based study. Cancer Epidemiol Biomarkers Prev. 2006;15(7):1331–5.
91. Li H, Stampfer MJ, Mucci L, Rifai N, Qiu W, Kurth T, et al. A 25-year prospective study of plasma adiponectin and leptin concentrations and prostate cancer risk and survival. Clin Chem. 2010;56(1):34–43.
92. Freedland SJ, Sokoll LJ, Mangold LA, Bruzek DJ, Mohr P, Yiu SK, et al. Serum leptin and pathological findings at the time of radical prostatectomy. J Urol. 2005;173(3):773–6.
93. Frankenberry KA, Somasundar P, McFadden DW, Vona-Davis LC. Leptin induces cell migration and the expression of growth factors in human prostate cancer cells. Am J Surg. 2004;188(5):560–5.
94. van Roermund JG, Hinnen KA, Tolman CJ, Bol GH, Witjes JA, Bosch JL, et al. Periprostatic fat correlates with tumour aggressiveness in prostate cancer patients. BJU Int. 2011;107(11):1775–9.
95. Simpson ER, Davis SR. Minireview: aromatase and the regulation of estrogen biosynthesis–some new perspectives. Endocrinology. 2001;142(11):4589–94.
96. Hsing AW, Chua Jr S, Gao YT, Gentzschein E, Chang L, Deng J, et al. Prostate cancer risk and serum levels of insulin and leptin: a population-based study. J Natl Cancer Inst. 2001;93(10):783–9.
97. Kasper JS, Liu Y, Giovannucci E. Diabetes mellitus and risk of prostate cancer in the health professionals follow-up study. Int J Cancer. 2009;124(6):1398–403.

Chapter 4
Increased Adiposity and Endometrial Cancer Risk

Karen H. Lu, Ann H. Klopp, Pamela T. Soliman, and Rosemarie E. Schmandt

Abstract Among all cancers, increasing body mass index is most strongly associated with endometrial cancer incidence and mortality. The molecular mechanisms underlying the role of obesity, and in particular the role of visceral fat, to the pathogenesis of endometrial cancer are becoming better understood. The current body of knowledge suggests several rational strategies, including behavioral, pharmaceutical, and surgical interventions, can be used to circumvent or derail the aberrant signaling pathways and hormonal abnormalities associated with obesity. Given the growing worldwide obesity epidemic, the development and availability of therapeutics, which can reduce the impact of obesity on endometrial cancer risk is imperative.

4.1 Introduction

Obesity increases the incidence of many cancers, such as breast, prostate, and colon cancer. However, endometrial cancer is the mostly tightly linked with obesity. Estimates suggest that nearly 40 % of cases of endometrial cancer can be attributed to obesity. Endometrial cancer is often divided into two biological entities, type I and type II. Type I endometrial cancer, which has a more favorable prognosis, is typically endometrioid, and associated with obesity and estrogen exposure. Type II endometrial cancer includes higher-risk subtypes such as uterine serous carcinoma, carcinosarcoma (malignant mixed mullerian tumor), and clear cell carcinoma. Type

K.H. Lu, M.D. (✉) • P.T. Soliman, M.D., M.P.H. • R.E. Schmandt, Ph.D.
Department of Gynecologic Oncology and Reproductive Medicine, The University of Texas MD Anderson Cancer Center, 1515 Holcombe Boulevard, Houston, TX 77030, USA
e-mail: khlu@mdanderson.org

A.H. Klopp, M.D., Ph.D.
Department of Radiation Oncology, The University of Texas MD Anderson Cancer Center, 1515 Holcombe Boulevard, Houston, TX 77030, USA

M.G. Kolonin (ed.), *Adipose Tissue and Cancer*, DOI 10.1007/978-1-4614-7660-3_4,

1 endometrial cancers are hormonally responsive and develop in an environment of excess exogenous or endogenous estrogen exposure [1, 2]. The most common cause of increased endogenous estrogen exposure is obesity due to biochemical changes in peripheral adipose tissue resulting in high serum estrogen and androgen levels with relatively lower progesterone levels. As a result, obesity and a sedentary lifestyle are strongly correlated with an increased incidence of endometrioid endometrial cancers. Other causes of increased endogenous estrogen exposure include estrogen-secreting tumors, low parity, early menarche, and late menopause [3, 4].

In addition to increased endogenous estrogen exposure, many other mechanisms have been proposed to account for the relationship between obesity and endometrial cancer, such as systemic and paracrine signaling through hormonal, growth factor, or metabolic pathways. Adipose tissue may also serve as a reservoir for tumor-supporting stromal cells [5]. To combat these mechanisms, preventative and therapeutic approaches are under investigation to disrupt these signaling pathways. The optimal approach for behavioral strategies to reduce adipose tissue, such as diet and exercise, are also under active investigation.

4.2 Epidemiology

4.2.1 Endometrial Cancer Incidence

Body Mass Index

Obesity is a strong risk factor for developing endometrial cancer. Obese women have a threefold higher risk of developing endometrial cancer than lean women [6]. In fact, every increase in BMI of 5 kg/m^2 increases a woman's risk of the developing of endometrial cancer by approximately 60 % (relative risk, 1.59; 95 % confidence interval [CI], 1.50–1.68) [7]. Endometrial cancer in obese women is more likely to have lower risk features such as endometrioid histology and low/intermediate grade.

Visceral Adiposity

The anatomic distribution of adipose tissue also impacts the risk of endometrial, ovarian, and other intra-abdominal cancers [8]. The incidence of colon, pancreatic, and prostate cancer is increased in individuals with excess of visceral fat [9]. An elevated waist-to-hip ratio, reflecting a preferential deposition of adipose in the abdomen, increases the risk of developing endometrial cancer by 220 % [10].

The cancer-promoting effects of excess visceral adipose may be due to differences in visceral and subcutaneous adipose such as increased regional and systemic secretion of inflammatory factors. Alternatively, the proximity of visceral adipose to the uterus may account for the higher risk of endometrial cancer in women with excess visceral adiposity. The primary site of intra-abdominal or visceral adipose tissue is in the omentum, which is a layer of fibrovascular fatty tissue, which surrounds the bowel. Omental adipose tissue is structurally and functionally distinct from subcutaneous adipose tissue and an excess of intra-abdominal adipose tissue is associated with higher rates of diverse disease states such as cardiovascular disease and diabetes [11].

4.2.2 Endometrial Cancer Mortality

Among the population as a whole, obesity increases the risk of death from endometrial cancer. In a study of 900,000 prospectively followed healthy patients, 57,145 individuals died of cancer over 16 years. The relative risk of death from endometrial cancer in this population was 6.25 for women with a BMI >40 and 2.77 with a BMI between 35 and 39 [6].

Teasing out the impact of obesity after treatment for endometrial cancer is difficult given the correlation with obesity and more indolent risk features such as lower grade and endometrioid histology. A recently reported meta-analysis found that while all-cause mortality was increased after diagnosis of endometrial cancer, there was no evidence supporting a role for obesity in cancer-specific mortality after treatment [12]. These studies attempted to take into account other risk factors that may be correlated with obesity to determine the direct impact of obesity on cancer survival. Recently, the impact of obesity on survival after endometrial cancer treatment was tested in a cohort of patients treated in a large randomized study, the MRC ASTEC study. Patients were divided into type 1 and type 2 endometrial cancers. Multivariate models including clinical risk factors found no evidence for worse survival among obese patients [for type 1 endometrial cancer the HR was 0.98 (95 % CI 0.86, 1.13)]. The impact of visceral obesity on survival after treatment for endometrial cancer remains unknown.

4.3 Mechanisms Through Which Adiposity Promotes Endometrial Cancer Initiation and Progression

4.3.1 Estrogen

The influence of unopposed estrogen on endometrial cancer risk is underscored by the observation that among women on unopposed exogenous estrogen, 20–50 % developed endometrial hyperplasia within 1 year [13–15]. In normal, premenopausal women, estrogen is produced predominantly by the granulosa cells of the ovaries and is a regulator of the normal reproductive cycle, promoting the cyclic renewal of endometrial tissue [16]. However, in postmenopausal women, peripheral tissues, and in particular adipose tissues, become the major sites of estrogen synthesis. Due to the increased activity of aromatase in adipose stem cells and the stromal cells of subcutaneous and abdominal fat, androgens produced by the adrenal glands are locally converted to estrone and estradiol. The increased activity of aromatase in adipose stem cells and the stromal cells of subcutaneous and abdominal fat results in the conversion of androgens produced by the adrenal glands to estrone and estradiol resulting in a "hyperestrogenic" state.

Aromatase levels in postmenopausal women increase both with adiposity and age [17–20]. Circulating levels of estrogens have been shown to increase with BMI, while adipose tissue concentrations of estrogen have been measured at levels several times higher that those observed in plasma [21]. Furthermore, obese women are also more likely to have other endocrine abnormalities that lead to low circulating level of sex hormone-binding globulin (SHBG) resulting in increased steroid hormone bioavailability, further increasing the endometrial exposure to circulating estrogen [22, 23].

While bioactive local and circulating estrogen increase an obese woman's risk for endometrial cancer, it is but one of many obesity-related factors that increase risk for endometrial cancer.

4.3.2 *Adipokines*

Adipose tissue secretes a number of metabolically active cytokines and hormones including adiponectin, leptin, resistin, and tumor necrosis factor-alpha (TNFα). Adiponectin, the most abundant adipokine, is secreted exclusively by adipocytes and stimulates the sensitivity of peripheral tissues to insulin. Low levels of adiponectin have been shown to have a high correlation with hyperinsulinemia and the degree of insulin resistance independent of adiposity. Adiponectin is also decreased in obesity as well as type 2 diabetes, which is generally thought to be an independent risk factor for endometrial cancer.

Several case–control studies have evaluated the association between serum adiponectin and endometrial cancer at the time of cancer diagnosis. The first study by Petridou and colleagues evaluated 84 cases and control and found that an increase in one standard deviation of adiponectin was associated with more than a 50 % reduction in the risk for endometrial cancer among women under 65, even after controlling for BMI [24]. A second study by Dal Maso and colleagues compared serum levels of adiponectin in 87 cases and 132 controls. They found that BMI and adiponectin level were independently associated with endometrial cancer [25]. Women with higher levels of adiponectin, those considered insulin sensitive, were less likely to have endometrial cancer. Finally, Soliman et al. showed the association between adiponectin and endometrial cancer was not only independent of BMI but also more profound than the association between BMI and endometrial cancer [26]. Each of these studies supports an inverse association of adiponectin levels and endometrial cancer at the time of diagnosis.

There have been two published prospective studies evaluating prediagnostic serum adiponectin and the risk of developing endometrial cancer. The first was published by Cust et al. in 2007 [27]. This was a case–control study nested within the European Prospective Investigation into Cancer and Nutrition. In this prospective study, 284 women who were not taking hormone replacement therapy developed endometrial cancer during the 5-year follow-up time. 548 controls were selected matched for age, menopause status, phase of menstrual cycle, time of blood draw, and fasting status. They found that women with the highest level of adiponectin at baseline were significantly less likely to develop endometrial cancer with nearly a 50 % reduction in risk compared to women with the lowest levels of adiponectin. A similar prospective case–control study nested within the Nurses Health Study was performed [26]. Among the 32,000 women enrolled in the study, 146 cases of histologically confirmed endometrioid adenocarcinoma were identified. Controls were randomly matched based on age, menopausal status, use of hormone replacement therapy, and fasting status at the time of blood draw. The median interval of time between blood draw and cancer diagnosis was 7 years with a range of 2–13 years. Baseline adiponectin level was not a statistically significant predictor of endometrial cancer risk. Additional studies are being performed to determine if insulin resistance as measured by adiponectin is a modifiable risk factor for the development of endometrial cancer.

4.3.3 *Inflammation*

Inflammatory diseases have long been linked to an increased risk of a variety of cancers. In fact, inflammation has recently been classified as a "tumor-enabling" hallmark of cancer [28]. Increasing abdominal adiposity (visceral or belly fat) is associated with a chronic state of inflammation. Not only are proinflammatory adipokines produced by adipocytes but like solid tumors, visceral adipose tissues also are infiltrated by proinflammatory immune cells including macrophages, T and B cells [29–32].

As a component of adipose tissue in obese individuals, immune cells, and specifically macrophages, secrete a variety of growth, survival, and proangiogenic factors, as well as bioactive molecules that enable tumor growth and contribute to the remodeling of the tumor microenvironment to facilitate metastases. Furthermore, reactive oxygen and nitrogen species released by activated macrophages are mutagenic and accelerate oncogenic mutations that contribute to cancer risk and progression [30, 33]. So, not only does inflamed visceral adipose tissue provide an ideal milieu for the growth of metastatic endometrial cancer but proinflammatory factors also secreted by infiltrating adipose immune cells mediate systemic effects on tumor progression at distant sites, including the endometrium.

Key growth factors secreted by macrophages, which affect tumor growth include TNFα and interleukin-6 (IL-6). Elevated levels of TNFα and IL-6 are observed in obese versus lean individuals [34, 35]. The TNFα receptor, TNF-R1 is broadly expressed in normal and tumor tissues including endometrial tumors. In normal premenopausal endometrium, TNFα plays a role in endometrial shedding and regeneration [36]. A case–control study conducted within the European prospective investigation into cancer and nutrition (EPIC) revealed that prediagnostic levels of circulating TNFα and its soluble receptors were related to a higher risk of endometrial cancer [4].

TNFα has been shown to play a role in both tumor initiation and tumor progression. TNFα signaling through TNFR1 activates the downstream NFkB transcription factor, which promotes the expression of proteins involved in inflammation, proliferation, survival, invasion, and metastasis [37]. TNF-R1 further mediates mitotic and cell survival signaling through the proproliferative AKT and MAPK signaling pathways.

The cytokine IL-6 is known to have both pro- and anti-inflammatory properties. In the context of obesity, IL-6 is secreted by T cells and macrophages infiltrating omental fat, as well as by adipocytes. Normal premenopausal endometrial tissue expresses both IL-6 and the IL-6R, and their levels increase upon transition from the proliferative to the secretory phase, coincidental with the implantation window [38]. In fact, IL-6 plays a role in facilitating embryo implantation, placental development and is key to the immune tolerance of pregnancy [39]. IL-6 has also been shown to participate in the regulation of stem cell renewal and is an important component of the inflammation-induced healing process [40, 41]. In breast stem cell mammospheres, Sansone et al. demonstrated that signaling through the IL-6 receptor induces the Jak/STAT-3 and MAPK pathways [40]. IL-6 further upregulates the expression of Jagged 1, and which amplifies signaling through its receptor, Notch3 and promotes the development of a more malignant mammosphere phenotype. While this effect has not been formally investigated in the transformation of normal endometrium, the expression of the IL-6R by normal endometrium and by endometrial cancer cells suggest this may represent a potential mechanism involved in the development and progression of endometrial cancer.

Like TNFα, a significant increase in endometrial cancer risk is associated with increasing serum concentrations of IL-6 [4]. A direct role for IL-6 in the postmenopausal endometrium has not yet been defined; however, one possible mechanism by which IL-6 contributes to the pathogenesis of endometrial cancer is its postulated role in aromatase synthesis by adipocytes [42]. Given that adipocytes are the primary site of estrogen synthesis in postmenopausal women, IL-6 produced by adipocyte-associated macrophages may promote estrogen synthesis by adipocytes and drive endometrial hyperplasia and ER-positive tumor growth.

Taken together, the expression of TNFα, IL-6, and other inflammatory cytokines produced by infiltrating macrophages as a consequence of abdominal adiposity provides another plausible mechanism linking obesity and endometrial cancer risk and progression.

4.3.4 *Insulin/IGF-1*

The chronic inflammation associated with abdominal obesity plays a critical role in the pathogenesis of insulin resistance, a condition that can ultimately lead to type 2 diabetes [43], a disease closely linked to endometrial cancer. TNFα was first described to inhibit insulin receptor signaling by inducing phosphorylation of the IRS-1 [44], thereby reducing the cellular response to insulin. Although less well defined, IL-6, leptin, adiponectin, and resistin are thought to further modulate insulin sensitivity [43].

The reduction in glucose uptake, which occurs as a consequence of insulin resistance, leads to hyperglycemia and the compensatory secretion of even more insulin by the pancreas. Ever-increasing levels of circulating insulin causes a reduction in insulin-like growth factor binding protein (IGF-BPs) synthesis by the liver and endometrium, leading to the increased bioavailability of circulating insulin-like growth factors (IGF-1 and 2). Together, insulin and IGFs signaling through the insulin-receptor (IR) and IGF-R promote cellular proliferation and prevent apoptosis through both the MAPK and PI3K signaling pathways.

IGF-1 promotes the proliferation of normal endometrial tissue [45]. An increase in IGF-1R is observed in endometrial hyperplasia, the precursor lesion to endometrial cancer, and heightens the responsiveness of these cells to IGF-1 [46]. Due in part to an increase in IGF-1-mediated signaling, hyperactivity of the PI3K/AKT/mTOR pathway is frequently observed in endometrial cancer. Compounding this proliferative signal is the concomitant loss of the phosphatase and tensin homolog (PTEN) lipid/protein phosphatase tumor suppressor gene, which is observed in greater than 40 % of endometrial cancers [47]. Under normal conditions, PTEN dephosphorylates PI3K substrates and acts as a PI3K antagonist and tumor suppressor gene. Together, the loss of PTEN coupled with elevated circulating IGF1 facilitates proproliferative signaling, which fuels the progression of endometrial proliferation and the progression to endometrial hyperplasia and cancer.

4.3.5 Cellular Metabolism

The characterization of metabolic changes that are observed in normal versus tumor tissues as a consequence of increased caloric intake and obesity has become an active area of research in the field of cancer prevention as a whole. Curiously, despite energy excess, human tumor cells rely on aerobic glycolysis and lipolysis for energy production [28, 48–50]. In fact, the use of aerobic glycolysis by tumor cells has been exploited for tumor-imaging purposes in the clinic using FDG-PET. Tumors preferentially take up 2-deoxy-2-[18F]fluoro-D-glucose [51], which is phosphorylated by hexokinase to produce FDG-6-phosphate. This isotopically labeled glucose cannot be further metabolized and can be used to determine relative levels of glucose uptake in tumors and allow the tumor to be visualized by positron emission tomography. Endometrial cancer typically demonstrates a high 2-deoxy-2-[18F]fluoro-D-glucose phenotype [48], indicating a metabolically active tumor type.

The hyperglycemia that accompanies obesity and insulin resistance therefore provides an obvious fuel source for metabolically active tumors such as endometrial cancer [52]. Adipocytes are further thought to supply tumors with fatty acids that further accelerate tumor growth. Of gynecologic tumors, both serous ovarian adenocarcinoma and uterine papillary serous carcinoma (type II endometrial cancer) metastasize preferentially to the omentum where they have easy access this fuel source [53, 54]. Hypertrophic adipocytes are further associated the release of free fatty acids (FFAs) into circulation and with aberrant trafficking of fatty acids to nonadipose tissue [55, 56]. This suggests that even early stage endometrial tumors, distant from the omentum, may be influenced by abdominal obesity.

4.3.6 Adipose Stem Cells

The tumor-promoting effect of excess fat near the tumor may be explained by the recent appreciation of a tumor-tropic adipose stem cell (ASC) population, which is harbored in adipose tissue [57]. These cells resemble the more well-described bone marrow derived mesenchymal stem cells [58, 59]. MSC and ASC are unique among mesenchymal cell populations in that they are highly migratory with the capacity to home to tumors. Once engrafted in tumors, they form tumor-associated fibroblasts (TAF). MSC/ASC can be recruited to form (TAFs) from both regional and distant tissues including circulating bone marrow derived cells and adipose tissue derived cells [60–64]. Recent work has demonstrated TAFs from distinct sources exhibit unique phenotypes with bone marrow derived cells expressing fibroblast-specific protein (FSP) and fibroblast activation protein (FAP), while pericytes, myofibroblasts, and endothelial cells appear to be derived from neighboring adipose tissue [64].

The impact of these cells on tumor biology remains the source of some debate [65]. In xenograft models, MSC can have dramatic effects on tumor progression, increasing tumor initiation and promoting metastasis [65–69]. Many studies have reported contradicting results; with some investigators finding that MSCs promote tumor growth and others reporting that MSCs inhibit tumor growth. Many mechanisms have been reported to account for these observations such as chemokine signaling, modulation of apoptosis, vascular support, and immune modulation. In this chapter, we analyzed the differences in the methodology of the studies reported and found that the timing of MSC introduction into tumors may be a critical element.

Of particular relevance to endometrial cancer, Klopp et al. have recently identified a population of ASC from visceral adipose tissue [70]. ASC isolated from the omentum (O-ASC1 and 2) of two patients with endometrial and ovarian cancer, morphologically and phenotypically resembled bone marrow-derived MSC (BM-MSC) and subcutaneous adipose derived MSC (SC-MSC). O-ASC shared cell surface marker expression and in vitro differentiation potential with BM-MCS and SC-ASC [70]. Tumor proliferation was monitored with luciferase imaging in the presence and absence of stromal cells. Growth of endometrial cancer cells in the presence of O-ASC2 significantly increased proliferation. Mice with Hec1A xenografts in the flank were injected into the hindlimb with O-ASC or SC-ASC, BM-MSC, and fibroblasts for comparison. In this model, the mesenchymal cells are remote from the xenograft tumor and must thus be recruited into the tumors to mimic the relationship of human malignancies and regional adipose tissue. Hec1A xenografts grew most rapidly when O-ASC were injected in the mouse hindquarter, demonstrating that visceral adipose tissue-derived stem cells isolated from patients with endometrial cancer have particularly potent effects on tumor progression. GFP-labeled O-ASC1 and O-ASC2 cells were detected in the peripheral stroma of the Hec1A xenografts demonstrating that O-ASC, like sc-ASC and BM-MSC, but not differentiated fibroblasts, exhibit tumor tropism [70].

4.4 Interventions for Cancer Prevention

As the molecular mechanisms underlying the contribution of obesity to cancer risk are identified, they suggest several rational strategies including behavioral, surgical, and pharmaceutical interventions, which can be exploited for endometrial cancer prevention.

4.4.1 Behavioral Interventions: Diet and Physical Activity

Weight loss and physical activity can reverse many of the negative molecular changes associated with obesity. A modest weight loss of 5–10 % of body weight has been shown to decrease incidence of type 2 diabetes and other obesity associated comorbidities and is currently recommended as a weight loss goal by the American Cancer Society for overweight cancer survivors [71–73]. Indeed, BMI and level of physical activity are modifiable behaviors that are associated with a decrease endometrial cancer risk. The American Cancer Society Cancer Prevention Study II Nutrition Cohort evaluated 42,672 postmenopausal women over an 11-year period with respect to BMI and physical activity in relation to the incidence of endometrial cancer [71–73]. Moderate levels of physical activity were associated with a 33 % lower risk of endometrial cancer, and the effect was most pronounced among overweight and obese women. These results and others suggest that 70% of endometrial cancers could be prevented by maintaining a healthy weight and regular exersize (Fig. 4.1).

Caloric restriction or increased physical activity is known to increase insulin sensitivity and is associated with a decrease in bioavailable IGF1, resulting from increased IGFBP synthesis. Further, weight loss is associated with a decrease in circulating adipokines. While detailed studies in endometrial cancer are lacking, transcriptional profiling performed on rectosigmoid mucosal biopsies from obese women, both pre- and postdiet induced 10 % weight loss revealed a decrease in a variety of adipokines including TNFα and IL-6, reduced tissue infiltration by T cell and macrophages as compared to preweight loss samples. These changes were reflected by a decrease in circulating biomarkers such as inflammatory cytokines, glucose, and lipid profiles.

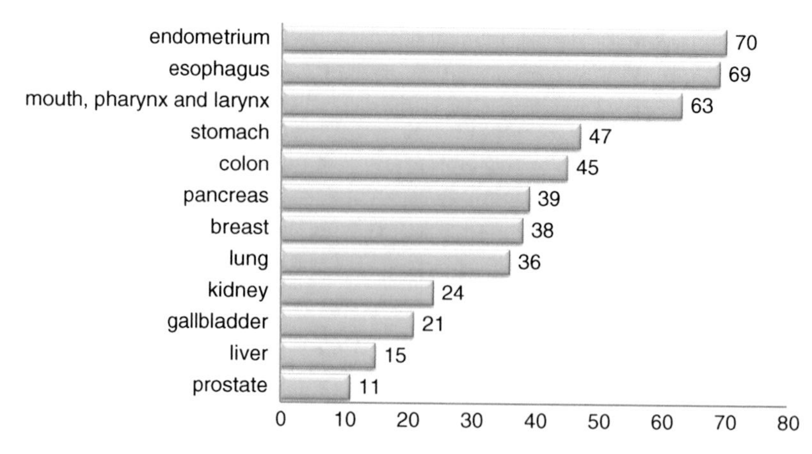

Fig. 4.1 Estimated percentage of cancers preventable in the USA with a healthy weight, diet, and regular physical activity. This material has been adapted from the 2009 WCRF/AICR Report *Policy and Action for Cancer Prevention. Food, Nutrition, Physical Activity and the Prevention of Cancer: a Global Perspective.* http://www.aicr.org/research/research_science_policy_report.html

Caloric restriction and exercise also activate AMPK (5′ adenosine monophosphate-activated protein kinase) as a consequence of increased AMP secondary to ATP depletion. The AMPK pathway represents an alternative pathway that works in opposition to PI3K/AKT/mTOR signaling. AMPK serves as the virtual "fuel gauge" of the cell and is activated in response to elevated AMP-to-ATP ratio in the cell that can occur under metabolic stress. The net effect of AMPK activation is the inhibition of energy-consuming processes (protein and lipid synthesis) and the activation of energy-producing processes (glycolysis) and a cessation of cellular, including tumor, proliferation.

Finally, weight loss and healthy food choices are also associated with decreasing levels of circulating estrogen in postmenopausal women. Studies performed to evaluate the effects dietary composition on breast cancer risk have circulating levels of bioactive estrogen can further be decreased with a high-fiber and low-fat diet, suggesting that dietary modifications, even in the absence of weight loss, may also be effective in reducing endometrial cancer risk.

Taken together, current evidence suggests that through a variety of mechanisms, weight loss and physical activity reduce proproliferative signaling and counteract environmental conditions that support the initiation and progression of endometrial cancer.

4.4.2 Pharmaceutical Intervention

The ability to sustain weight loss by diet and exercise requires significant lifestyle changes and has proven to be a challenge for many individuals. Surgical and pharmacologic options may need to be considered for the prevention of obesity-associated tumor types.

Contraceptives

In normal premenopausal endometrium, a balance between estrogen and progesterone are tightly maintained throughout the menstrual cycle. While estrogen drives the early proliferation of the endometrial epithelium, progesterone is growth inhibitory and promotes endometrial differentiation later in the cycle. Physiologic conditions producing a hormonal imbalance favoring an excess of estrogen or a suppression of progesterone synthesis support the development of endometrial hyperplasia and increased endometrial cancer risk. This indicates that the proproliferative hyperestrogenic milieu, such as that found in the obese, postmenopausal women, may be normalized using progesterone-based therapy and would therefore be protective against endometrial cancer [74–76].

Synthetic progestins are commonly used in oral contraceptives and intrauterine devices and for the treatment of menorrhagia, infertility, and the symptoms of menopause. Progestin therapy is also standard for the treatment of women who have endometrial hyperplasia without atypia, the precursor lesion for endometrial cancer [77]. The response of complex hyperplasia with atypia and well-differentiated endometrial carcinoma to progestin has also been investigated. In studies performed by Wheeler et al., complete remission was observed in 67 % of women with complex atypical hyperplasia, while 11 % regressed to complex hyperplasia without atypia, and 22 % had persistent disease with an 11-month follow-up [77]. In women with well-differentiated endometrial carcinoma, 42 % of women demonstrated a complete remission, while 58 % had persistent disease over a 12-month follow-up period. Multiple epidemiologic studies demonstrate that women who use combination estrogen and progesterone oral contraceptives (OCP) decrease their risk of endometrial cancer by 50 % [78–80]. While there is no data to support a decreased efficacy in endometrial cancer protection in obese women, there are studies that suggest that obese women have a slightly decreased contraceptive efficacy compared to thin women [81]. A study by Maxwell et al. found that OCP with higher potency progestins may be more effective for women with a higher BMI [82]. More data will be necessary to refine the optimal OCP for endometrial cancer chemoprevention in an obese population.

Recently, the levonorgestrel containing IUD (LNG-IUD) has become an attractive candidate for endometrial cancer prevention. The IUD alone has been shown to decrease endometrial cancer risk, and the progestin may further protect the endometrium [83]. A few studies have reported the use of the LNG-IUD for reversal of endometrial hyperplasia and cancer, but additional data will be necessary before it can be recommended for routine endometrial cancer prevention [81, 84, 85].

Metformin (Antidiabetic Agents)

Metformin is an oral antihyperglycemic drug long used in the management of type 2 diabetes [86]. Metformin lowers blood glucose levels by inhibiting gluconeogenesis and improves insulin sensitivity by increasing peripheral glucose uptake and utilization. The observation that type 2 diabetics being treated with metformin had a reduced risk of cancer has sparked a flurry of research investigating the value of this drug as an antineoplastic agent in a variety of tumor types [87].

Based on its positive systemic effect on increasing insulin sensitivity, and the close association between hyperinsulinemia, type 2 diabetes and endometrial cancer risk, metformin represents a rational chemopreventive agent for endometrial cancer. Additional studies suggest the chemopreventive and therapeutic benefits of metformin are mediated directly through activation of the growth inhibitory AMPK pathway [87] and inhibit the PI3K/AKT/mTOR pathway hyperactivity observed in a majority of endometrial cancers. In support of this mechanism, our group and others have demonstrated a clear and direct antiproliferative effect of metformin on a variety of endometrial cancer cell lines [88–91].

The mechanisms by which metformin exerts its antiproliferative effects have not been fully elucidated, and this drug has demonstrated several additional direct effects, which predict its utility the prevention and treatment of endometrial cancer. Recently, metformin has been shown to inhibit aromatase expression by primary cultures of human adipose stromal cells [92]. It is likely, therefore, that metformin can inhibit the localized production of estrogens in tumor stromal tissue and also decrease circulating levels of estrogen observed in obese individuals. Metformin has also been shown to increase progesterone receptor expression and to reverse progestin resistance in endometrial cancer cell lines thereby potentially amplifying the antiproliferative effects of progesterone in endometrial tumors [93, 94].

These mechanisms of action suggest a promising and novel role for this "old drug" in the chemoprevention of a variety of primary and recurrent tumors [95, 96], and in particular, metformin represents an ideal drug choice for the primary and tertiary prevention of endometrial cancer. Clinical trials are currently underway to evaluate the use of metformin for endometrial cancer prevention in obese women.

4.4.3 Surgical Intervention: Bariatric Surgery

Recent studies have demonstrated a rapid improvement in the glycemic control, hyperinsulinemia, HOMA-IR index (homeostasis model assessment-estimated insulin resistance), or diabetes remission in obese patients following gastric bypass surgery as compared to conventional medical therapy [97, 98]. Furthermore, the amelioration of hyperglycemia and type 2 diabetes following bariatric surgery are sustained [99–101]. Given the tight association between obesity, insulin resistance, and endometrial cancer risk, and proposed role of insulin and IGF-1 in the pathogenesis of endometrial cancer, bariatric surgery represents a rational strategy for endometrial cancer prevention.

Indeed, by prolonging weight loss and preventing insulin resistance, several studies have recently shown that bariatric surgery can reduce both cancer risk and recurrence [101–103]. A 10–20-year follow-up of individuals in Sweden who underwent bariatric surgery demonstrated an overall reduced cancer risk as compared to obese individuals. Interestingly, this effect favored women (women: HR=0·58; $p=0.0008$; men: n.s) [101, 102]. A larger study of 6,596 US subjects who underwent gastric bypass surgery demonstrated a decreased incidence of cancer (HR 0.76, 95 % CI 0.65–89, $p=0.006$) [103]. In agreement with the Swedish study, bariatric surgery produced a more pronounced cancer-preventive effect in women (HR 0.73, 95 % CI 0.62–0.87, $p=0.0004$) over men (HR 1.02, 0.69–1.52, $p=0.91$). In the specific case of endometrial cancer risk, the hazard ratio was dramatically reduced by gastric bypass surgery to 0.22 (95 % CI 0.13–0.40, $p<0.0001$).

Data obtained from these surgical studies underscore the profound effect of obesity on endometrial cancer risk and suggest that weight loss, whether by behavioral change or surgical intervention, is critical to endometrial cancer prevention.

4.5 Future Directions

The growing appreciation of the mechanisms that account for the higher risk of endometrial cancer in obese women has opened the door to a number of strategies to prevent or treat endometrial cancer. Ongoing studies are focused on identifying obese women at the highest risk of developing endometrial cancer, based on adipokine profile, adipose distribution, or changes in the endometrium. These women can then be selected for the optimal intervention, which may range from obesity prevention via behavioral or surgical approaches, pharmacologic strategies, or uterine-targeted therapy such as levonorgestrel containing IUD or hysterectomy. Similarly, investigations are underway to assess the benefit of diet and exercise, metformin, and biologically targeted agents to disrupt signaling via inflammatory adipose derived cytokines to treat endometrial cancer. We anticipate these approaches will prove effective in reducing the rising incidence and mortality from endometrial cancer.

References

1. Fader AN et al. Endometrial cancer and obesity: epidemiology, biomarkers, prevention and survivorship. Gynecol Oncol. 2009;114(1):121–7.
2. Schouten LJ, Goldbohm RA, van den Brandt PA. Anthropometry, physical activity, and endometrial cancer risk: results from the Netherlands Cohort Study. J Natl Cancer Inst. 2004; 96(21):1635–8.
3. Brinton LA et al. Reproductive, menstrual, and medical risk factors for endometrial cancer: results from a case-control study. Am J Obstet Gynecol. 1992;167(5):1317–25.
4. Dossus L et al. Reproductive risk factors and endometrial cancer: the European Prospective Investigation into Cancer and Nutrition. Int J Cancer. 2010;127(2):442–51.
5. Zhang Y et al. White adipose tissue cells are recruited by experimental tumors and promote cancer progression in mouse models. Cancer Res. 2009;69(12):5259–66.
6. Calle EE et al. Overweight, obesity, and mortality from cancer in a prospectively studied cohort of U.S. adults. N Engl J Med. 2003;348(17):1625–38.

7. Renehan AG et al. Incident cancer burden attributable to excess body mass index in 30 European countries. Int J Cancer. 2010;126(3):692–702.
8. Delort L et al. Central adiposity as a major risk factor of ovarian cancer. Anticancer Res. 2009;29(12):5229–34.
9. Balentine CJ et al. Intra-abdominal fat predicts survival in pancreatic cancer. J Gastrointest Surg. 2010;14(11):1832–7.
10. Friedenreich C et al. Anthropometric factors and risk of endometrial cancer: the European prospective investigation into cancer and nutrition. Cancer Causes Control. 2007;18(4): 399–413.
11. Ibrahim MM. Subcutaneous and visceral adipose tissue: structural and functional differences. Obes Rev. 2010;11(1):11–8.
12. Arem H, Irwin ML. Obesity and endometrial cancer survival: a systematic review. Int J Obes (Lond). 2012.
13. Furness S et al. Hormone therapy in postmenopausal women and risk of endometrial hyperplasia. Cochrane Database Syst Rev. 2009;2:CD000402.
14. Schiff I et al. Endometrial hyperplasia in women on cyclic or continuous estrogen regimens. Fertil Steril. 1982;37(1):79–82.
15. Woodruff JD, Pickar JH. Incidence of endometrial hyperplasia in postmenopausal women taking conjugated estrogens (Premarin) with medroxyprogesterone acetate or conjugated estrogens alone. The Menopause Study Group. Am J Obstet Gynecol. 1994;170(5 Pt 1): 1213–23.
16. Mihm M, Gangooly S, Muttukrishna S. The normal menstrual cycle in women. Anim Reprod Sci. 2011;124(3–4):229–36.
17. Brown KA, Simpson ER. Obesity and breast cancer: mechanisms and therapeutic implications. Front Biosci (Elite Ed). 2012;4:2515–24.
18. Cauley JA et al. The epidemiology of serum sex hormones in postmenopausal women. Am J Epidemiol. 1989;129(6):1120–31.
19. Simpson ER, Mendelson CR. Effect of aging and obesity on aromatase activity of human adipose cells. Am J Clin Nutr. 1987;45(1 Suppl):290–5.
20. Stocco C. Tissue physiology and pathology of aromatase. Steroids. 2012;77(1–2):27–35.
21. Szymczak J et al. Concentration of sex steroids in adipose tissue after menopause. Steroids. 1998;63(5–6):319–21.
22. Le TN et al. Sex hormone-binding globulin and type 2 diabetes mellitus. Trends Endocrinol Metab. 2012;23(1):32–40.
23. Potischman N et al. Case-control study of endogenous steroid hormones and endometrial cancer. J Natl Cancer Inst. 1996;88(16):1127–35.
24. Petridou E et al. Plasma adiponectin concentrations in relation to endometrial cancer: a case-control study in Greece. J Clin Endocrinol Metab. 2003;88(3):993–7.
25. Dal Maso L et al. Circulating adiponectin and endometrial cancer risk. J Clin Endocrinol Metab. 2004;89(3):1160–3.
26. Soliman PT et al. Circulating adiponectin levels and risk of endometrial cancer: the prospective Nurses' Health Study. Am J Obstet Gynecol. 2011;204(2):167.e1–5.
27. Cust AE et al. Plasma adiponectin levels and endometrial cancer risk in pre- and postmenopausal women. J Clin Endocrinol Metab. 2007;92(1):255–63.
28. Hanahan D, Weinberg RA. Hallmarks of cancer: the next generation. Cell. 2011;144(5): 646–74.
29. Chawla A, Nguyen KD, Goh YP. Macrophage-mediated inflammation in metabolic disease. Nat Rev Immunol. 2011;11(11):738–49.
30. Qian BZ, Pollard JW. Macrophage diversity enhances tumor progression and metastasis. Cell. 2010;141(1):39–51.
31. Schipper HS et al. Adipose tissue-resident immune cells: key players in immunometabolism. Trends Endocrinol Metab. 2012;23(8):407–15.
32. Sell H, Habich C, Eckel J. Adaptive immunity in obesity and insulin resistance. Nat Rev Endocrinol. 2012;8(12):709–16.

33. Mantovani A et al. Cancer-related inflammation. Nature. 2008;454(7203):436–44.
34. Kern PA et al. Adipose tissue tumor necrosis factor and interleukin-6 expression in human obesity and insulin resistance. Am J Physiol Endocrinol Metab. 2001;280(5):E745–51.
35. Park EJ et al. Dietary and genetic obesity promote liver inflammation and tumorigenesis by enhancing IL-6 and TNF expression. Cell. 2010;140(2):197–208.
36. Haider S, Knofler M. Human tumour necrosis factor: physiological and pathological roles in placenta and endometrium. Placenta. 2009;30(2):111–23.
37. Sethi G, Sung B, Aggarwal BB. TNF: a master switch for inflammation to cancer. Front Biosci. 2008;13:5094–107.
38. von Wolff M et al. Endometrial expression and secretion of interleukin-6 throughout the menstrual cycle. Gynecol Endocrinol. 2002;16(2):121–9.
39. Prins JR, Gomez-Lopez N, Robertson SA. Interleukin-6 in pregnancy and gestational disorders. J Reprod Immunol. 2012;95(1–2):1–14.
40. Sansone P et al. IL-6 triggers malignant features in mammospheres from human ductal breast carcinoma and normal mammary gland. J Clin Invest. 2007;117(12):3988–4002.
41. Schafer ZT, Brugge JS. IL-6 involvement in epithelial cancers. J Clin Invest. 2007;117(12): 3660–3.
42. Purohit A, Reed MJ. Regulation of estrogen synthesis in postmenopausal women. Steroids. 2002;67(12):979–83.
43. Donath MY, Shoelson SE. Type 2 diabetes as an inflammatory disease. Nat Rev Immunol. 2011;11(2):98–107.
44. Hotamisligil GS et al. IRS-1-mediated inhibition of insulin receptor tyrosine kinase activity in TNF-alpha- and obesity-induced insulin resistance. Science. 1996;271(5249):665–8.
45. Druckmann R, Rohr UD. IGF-1 in gynaecology and obstetrics: update 2002. Maturitas. 2002;41 Suppl 1:S65–83.
46. McCampbell AS et al. Overexpression of the insulin-like growth factor I receptor and activation of the AKT pathway in hyperplastic endometrium. Clin Cancer Res. 2006;12(21): 6373–8.
47. Hecht JL, Mutter GL. Molecular and pathologic aspects of endometrial carcinogenesis. J Clin Oncol. 2006;24(29):4783–91.
48. Bensinger SJ, Christofk HR. New aspects of the Warburg effect in cancer cell biology. Semin Cell Dev Biol. 2012;23(4):352–61.
49. Dang CV. Links between metabolism and cancer. Genes Dev. 2012;26(9):877–90.
50. Levine AJ, Puzio-Kuter AM. The control of the metabolic switch in cancers by oncogenes and tumor suppressor genes. Science. 2010;330(6009):1340–4.
51. Cubas R et al. Virus-like particle (VLP) lymphatic trafficking and immune response generation after immunization by different routes. J Immunother. 2009;32(2):118–28.
52. Becker S, Dossus L, Kaaks R. Obesity related hyperinsulinaemia and hyperglycaemia and cancer development. Arch Physiol Biochem. 2009;115(2):86–96.
53. Nieman KM et al. Adipocytes promote ovarian cancer metastasis and provide energy for rapid tumor growth. Nat Med. 2011;17(11):1498–503.
54. Zhang Y et al. Stromal progenitor cells from endogenous adipose tissue contribute to pericytes and adipocytes that populate the tumor microenvironment. Cancer Res. 2012;72(20): 5198–208.
55. Karpe F, Dickmann JR, Frayn KN. Fatty acids, obesity, and insulin resistance: time for a reevaluation. Diabetes. 2011;60(10):2441–9.
56. Mittendorfer B. Origins of metabolic complications in obesity: adipose tissue and free fatty acid trafficking. Curr Opin Clin Nutr Metab Care. 2011;14(6):535–41.
57. Muehlberg FL et al. Tissue-resident stem cells promote breast cancer growth and metastasis. Carcinogenesis. 2009;30(4):589–97.
58. Dominici M et al. Minimal criteria for defining multipotent mesenchymal stromal cells. The International Society for Cellular Therapy position statement. Cytotherapy. 2006;8(4):315–7.
59. Pereira RF et al. Cultured adherent cells from marrow can serve as long-lasting precursor cells for bone, cartilage, and lung in irradiated mice. Proc Natl Acad Sci USA. 1995;92(11): 4857–61.

60. Barcellos-Hoff MH. It takes a tissue to make a tumor: epigenetics, cancer and the microenvironment. J Mammary Gland Biol Neoplasia. 2001;6(2):213–21.
61. Klopp AH et al. Tumor irradiation increases the recruitment of circulating mesenchymal stem cells into the tumor microenvironment. Cancer Res. 2007;67(24):11687–95.
62. Studeny M et al. Mesenchymal stem cells: potential precursors for tumor stroma and targeted-delivery vehicles for anticancer agents. J Natl Cancer Inst. 2004;96(21):1593–603.
63. Studeny M et al. Bone marrow-derived mesenchymal stem cells as vehicles for interferon-beta delivery into tumors. Cancer Res. 2002;62(13):3603–8.
64. Kidd S et al. Origins of the tumor microenvironment: quantitative assessment of adipose-derived and bone marrow-derived stroma. PLoS One. 2012;7(2):e30563.
65. Klopp AH et al. Concise review: dissecting a discrepancy in the literature: do mesenchymal stem cells support or suppress tumor growth? Stem Cells. 2011;29(1):11–9.
66. Karnoub AE et al. Mesenchymal stem cells within tumour stroma promote breast cancer metastasis. Nature. 2007;449(7162):557–63.
67. Zhu W et al. Mesenchymal stem cells derived from bone marrow favor tumor cell growth in vivo. Exp Mol Pathol. 2006;80(3):267–74.
68. Djouad F et al. Immunosuppressive effect of mesenchymal stem cells favors tumor growth in allogeneic animals. Blood. 2003;102(10):3837–44.
69. Djouad F et al. Earlier onset of syngeneic tumors in the presence of mesenchymal stem cells. Transplantation. 2006;82(8):1060–6.
70. Klopp AH et al. Omental adipose tissue-derived stromal cells promote vascularization and growth of endometrial tumors. Clin Cancer Res. 2012;18(3):771–82.
71. Doyle C et al. Nutrition and physical activity during and after cancer treatment: an American Cancer Society guide for informed choices. CA Cancer J Clin. 2006;56(6):323–53.
72. Kushi LH et al. American Cancer Society Guidelines on Nutrition and Physical Activity for cancer prevention: reducing the risk of cancer with healthy food choices and physical activity. CA Cancer J Clin. 2006;56(5):254–81. quiz 313-4.
73. Kushi LH et al. American Cancer Society Guidelines on nutrition and physical activity for cancer prevention: reducing the risk of cancer with healthy food choices and physical activity. CA Cancer J Clin. 2012;62(1):30–67.
74. Yang S, Thiel KW, Leslie KK. Progesterone: the ultimate endometrial tumor suppressor. Trends Endocrinol Metab. 2011;22(4):145–52.
75. Yang S et al. Endometrial cancer: reviving progesterone therapy in the molecular age. Discov Med. 2011;12(64):205–12.
76. Carlson MJ et al. Catch it before it kills: progesterone, obesity, and the prevention of endometrial cancer. Discov Med. 2012;14(76):215–22.
77. Wheeler DT, Bristow RE, Kurman RJ. Histologic alterations in endometrial hyperplasia and well-differentiated carcinoma treated with progestins. Am J Surg Pathol. 2007;31(7):988–98.
78. Kaufman DW et al. Decreased risk of endometrial cancer among oral-contraceptive users. N Engl J Med. 1980;303(18):1045–7.
79. Beral V, Hannaford P, Kay C. Oral contraceptive use and malignancies of the genital tract. Results from the Royal College of General Practitioners' Oral Contraception Study. Lancet. 1988;2(8624):1331–5.
80. Combination oral contraceptive use and the risk of endometrial cancer. The Cancer and Steroid Hormone Study of the Centers for Disease Control and the National Institute of Child Health and Human Development. JAMA. 1987;257(6):796–800.
81. Lopez LM et al. Hormonal contraceptives for contraception in overweight or obese women. Cochrane Database Syst Rev. 2010;7:CD008452.
82. Maxwell GL et al. Progestin and estrogen potency of combination oral contraceptives and endometrial cancer risk. Gynecol Oncol. 2006;103(2):535–40.
83. Bahamondes L et al. Levonorgestrel-releasing intrauterine system (Mirena) as a therapy for endometrial hyperplasia and carcinoma. Acta Obstet Gynecol Scand. 2003;82(6):580–2.

84. Beining RM et al. Meta-analysis of intrauterine device use and risk of endometrial cancer. Ann Epidemiol. 2008;18(6):492–9.
85. Ismail MT, Fahmy DM, Elshmaa NS. Efficacy of levonorgestrel-releasing intrauterine system versus oral progestins in treatment of simple endometrial hyperplasia without atypia. Reprod Sci. 2013;20(1):45–50.
86. Ungar G, Freedman L, Shapiro SL. Pharmacological studies of a new oral hypoglycemic drug. Proc Soc Exp Biol Med. 1957;95(1):190–2.
87. Pollak MN. Investigating metformin for cancer prevention and treatment: the end of the beginning. Cancer Discov. 2012;2(9):778–90.
88. Mu N, Wang Y, Xue F. Metformin: a potential novel endometrial cancer therapy. Int J Gynecol Cancer. 2012;22(2):181.
89. Hanna RK et al. Metformin potentiates the effects of paclitaxel in endometrial cancer cells through inhibition of cell proliferation and modulation of the mTOR pathway. Gynecol Oncol. 2012;125(2):458–69.
90. Tan BK et al. Metformin treatment exerts antiinvasive and antimetastatic effects in human endometrial carcinoma cells. J Clin Endocrinol Metab. 2011;96(3):808–16.
91. Cantrell LA et al. Metformin is a potent inhibitor of endometrial cancer cell proliferation–implications for a novel treatment strategy. Gynecol Oncol. 2010;116(1):92–8.
92. Brown KA et al. Metformin inhibits aromatase expression in human breast adipose stromal cells via stimulation of AMP-activated protein kinase. Breast Cancer Res Treat. 2010;123(2): 591–6.
93. Shen ZQ, Zhu HT, Lin JF. Reverse of progestin-resistant atypical endometrial hyperplasia by metformin and oral contraceptives. Obstet Gynecol. 2008;112(2 Pt 2):465–7.
94. Xie Y et al. Metformin promotes progesterone receptor expression via inhibition of mammalian target of rapamycin (mTOR) in endometrial cancer cells. J Steroid Biochem Mol Biol. 2011;126(3–5):113–20.
95. Engelman JA, Cantley LC. Chemoprevention meets glucose control. Cancer Prev Res (Phila). 2010;3(9):1049–52.
96. Ben Sahra I et al. Metformin in cancer therapy: a new perspective for an old antidiabetic drug? Mol Cancer Ther. 2010;9(5):1092–9.
97. Schauer PR et al. Bariatric surgery versus intensive medical therapy in obese patients with diabetes. N Engl J Med. 2012;366(17):1567–76.
98. Mingrone G et al. Bariatric surgery versus conventional medical therapy for type 2 diabetes. N Engl J Med. 2012;366(17):1577–85.
99. Adams TD et al. Health benefits of gastric bypass surgery after 6 years. JAMA. 2012;308(11): 1122–31.
100. Heneghan HM et al. Effects of bariatric surgery on diabetic nephropathy after 5 years of follow-up. Surg Obes Relat Dis. 2013;9(1):7–14.
101. Sjostrom L. Review of the key results from the Swedish Obese Subjects (SOS) trial: a prospective controlled intervention study of bariatric surgery. J Intern Med. 2013;273(3): 219–34.
102. Sjostrom L et al. Effects of bariatric surgery on cancer incidence in obese patients in Sweden (Swedish Obese Subjects Study): a prospective, controlled intervention trial. Lancet Oncol. 2009;10(7):653–62.
103. Adams TD et al. Cancer incidence and mortality after gastric bypass surgery. Obesity (Silver Spring). 2009;17(4):796–802.

Chapter 5
Adipokines: Soluble Factors from Adipose Tissue Implicated in Cancer

Gilberto Paz-Filho, Ameet Kumar Mishra, and Julio Licinio

Abstract There is strong evidence for the association between obesity and cancer. Several retrospective and prospective observational studies have demonstrated that obesity and adiposity are independent risk factors for different types of cancer in both genders. According to a recent meta-analysis, a body mass index equal of higher than 40 kg/m^2 determines a relative risk for the development of all cancers equal to 1.52 for males and 1.88 for females.

The adipose tissues secretes several hormones, cytokines (named adipokines), inflammatory cytokines, factors related to complement and fibrinolysis, fatty acids, and enzymes. An increase in adiposity alters the homeostasis of those substances secreted by the adipose tissue and others (such as insulin and insulin-like growth factors). The pathophysiological bases of obesity-related cancer can be explained by alterations in adipokines levels, increase in insulin resistance, changes towards a proinflammatory state, and other effects such as increased oxidative stress.

Leptin and adiponectin are the most abundant adipokines, and both play a major role in the pathogenesis of obesity-related cancer. In this chapter, the role of adipokines in the pathogenesis of obesity-related cancer, with emphasis on leptin and adiponectin, is discussed.

G. Paz-Filho • A.K. Mishra
Department of Translational Medicine, The John Curtin School of Medical Research, The Australian National University, Canberra, ACT, Australia

J. Licinio (✉)
Mind and Brain Theme, South Australian Health and Medical Research Institute and Department of Psychiatry, Flinders University, Adelaide, South Australia
e-mail: julio.licinio@sahmri.com

M.G. Kolonin (ed.), *Adipose Tissue and Cancer*, DOI 10.1007/978-1-4614-7660-3_5,

5.1 Introduction

The prevalences of overweight (body mass index—BMI >25 kg/m^2 and <30 kg/m^2) and obesity (BMI≥30 kg/m^2) are rising to epidemic proportions worldwide and are major public concerns. In 2006, the World Health Organization (WHO) estimated that more than 1.4 billion adults (20 years and older) were overweight or obese. Combined, those conditions are the fifth leading cause of mortality. Each year, more than 2.8 million adult deaths are attributed to overweight or obesity [1]. Obesity leads to several comorbidities such as diabetes, dyslipidemia, hypertension, sleep apnea, osteoarthritis, menstrual disorders, infertility, gout, stroke, ischemic heart disease, congestive heart failure, deep vein thrombosis, and pulmonary embolism [2]. Overweight and obesity are attributable for 44 % of the diabetes burden, 23 % of the ischemic heart disease burden and between 7 % and 41 % of certain cancer burdens [1].

More recently, adipose tissue excess has been associated to an increased risk for the development of several types of cancer, as suggested by epidemiological studies and by meta-analyses. The pathophysiological basis for the effect of adiposity on the increase of cancer risk can be at least partially explained by the effects of adipokines—hormones that are synthesized and secreted by the white adipose tissue.

Here, we review the pathophysiological roles of adipokines in obesity-related cancer, emphasizing two of the most studied adipokines: leptin and adiponectin.

5.2 Epidemiology of Obesity-Related Cancer

The link between obesity and cancer risk has been studied for more than 50 years. The earliest and one of the largest epidemiological studies investigating the associations between weight and cancer was published in 1979. This was a long-term prospective study involving 750,000 men and women evaluated between 1959 and 1972, showing that mortality was approximately 90 % higher in men and women who were more than 40 % heavier than the average. In that group, cancer mortality was also increased by a third among men, due to colon and rectum cancers, and by 55 % among women, due to cancers of the gallbladder, biliary passages, breast, cervix, endometrium, and ovary [3]. The association between weight and cancer has also been supported by a report published by the International Agency for Research into Cancer (IARC) in 2005 [4]. In 2008, the World Cancer Research Fund issued an Expert Report acknowledging the association between excess body fat with increased risk of esophageal adenocarcinoma and cancers of the pancreas, colon, rectum, postmenopausal breast, endometrium, and kidney [5]. Subsequently, a very large epidemiological study, the Million Women Study, identified associations between high BMI and increased risk of endometrial cancer (2.39), esophageal adenocarcinoma (2.38), premenopausal colorectal cancer (1.61), kidney cancer (1.53), leukemia (1.50), postmenopausal breast cancer (1.40), multiple myeloma

Table 5.1 Relative cancer risk by BMI categories (95 % confidence interval)

Cancer types	BMI (kg/m²)	Male	Female
All cancers	≥40	1.52	1.88
Liver	≥35	4.52	1.68
Pancreas	≥35	2.61	2.76
Kidney	≥35 male, ≥40 female	1.70	4.75
Colon and rectum	≥35 male, ≥40 female	1.84	1.46
Esophagus	≥30	1.91	2.64
Gallbladder	≥30	1.76	2.13
Multiple myeloma	≥35	1.71	1.44
Non-Hodgkin's lymphoma	≥35	1.49	1.95
Prostate	≥35	1.34	NA
Stomach	≥35	1.94	Data N/a
Uterus	≥40	NA	6.25
Cervix	≥40	NA	3.20
Breast	≥40	NA	2.12
Ovary	≥35	NA	1.51
All other cancers	≥30 male, ≥40 female	1.68	2.51

NA not applicable, *N/a* not available
Adapted from [24]

(1.31), pancreatic cancer (1.24), non-Hodgkin's lymphoma (1.17), and ovarian cancer (1.14). The risk for all cancers combined was 1.12 [6].

The first meta-analysis published in 2001 demonstrated that nearly 35,000 and 37,000 new cases of cancer in Europe were related to obesity and to overweight, respectively. The cancers that were most strongly associated with overweight and obesity were endometrium, kidney, gallbladder, colon, and breast [3]. Several other meta-analyses have been published, associating obesity or high BMI with increased risk for several types of cancer: colon (both sexes) and rectal (men) [7–10], gallbladder [11], liver [12, 13], kidney [14, 15], pancreas [16, 17], ovaries [18, 19], breast [20], and prostate cancers [21]; leukemia, non-Hodgkin's lymphoma, and multiple myeloma [22, 23].

Based on a large population study, US Cancer Prevention Study II, obesity was linked to 20 % of all cancer deaths in women and 14 % in men [24]. In this study, 1,184,617 participants were selected and prospective mortality was studied for 16 years starting from 1982. The subgroups with cancer-related deaths were evaluated based on BMI to investigate relative risk of death of cancer because of obesity. The relative risk of death in different types of cancer for the highest BMI category both in men and women is expressed in Table 5.1 [24]. In another meta-analysis, the population attributable risk factor for incidence of cancer was 3.2 % in men and 8.6 % in women [25]. Additionally, a study in Swedish obese patients undergoing bariatric surgery for morbid obesity and their long-term follow-up indicated a reduction in cancer incidence [26], pointing causal association between obesity and cancer risk.

Table 5.2 Classes of secretory factors from the adipose tissue

Category	Molecules
Hormones and adipokines	Estrogen, leptin, adiponectin, resistin, angiotensinogen, visfatin
Inflammatory mediators	Interleukin-6 (IL-6), tumor necrosis factor-α (TNF-α), monocyte chemoattractant protein (MCP-1), macrophage migration inhibitory factor (MIF), interleukin-1 receptor antagonist (IL-1Ra)
Complement-related factors	Adipsin, complement factor B, acylation-stimulating protein (ASP)
Extracellular matrix	Type I, III, IV, and VI collagen, fibronectin, osteonectin, laminin, entactin, matrix metalloproteinase (MMP-2)
Fibrinolytic system	Plasminogen activator inhibitor-1 (PAI-1)
Others	Free fatty acids (FFA), lipoprotein lipase (LPL), prostacyclin, retinol binding protein 4, angiopoietin-like protein 4 (ANGPTL4)

5.3 Pathophysiology of Obesity

At the cellular level, overweight and obesity are characterized by the increase in number and size of adipocytes. A lean adult has 35 million adipocytes, each containing 0.4–0.6 μg of triglycerides, whereas an extremely obese person has 125 million adipocytes, each containing 0.8–1.2 μg of triglycerides [27]. Traditionally, adipocytes have been viewed solely as energy depots, but after the discovery of leptin in 1994 and extensive research in the field in the last decades, it has been established that the adipose tissue is an active endocrine organ.

The adipocyte is a major source of secreted proteins and accounts for 19.6 % of the total transcripts expressed in the adipose tissue [28]. Over 50 substances, called adipokines, have been identified to date and have range of physiological functions such as energy homeostasis, metabolism, immunity, and inflammation (Tables 5.2 and 5.3).

5.4 Pathophysiological Bases of Obesity-Related Cancer

The most widely studied mechanisms describing the underlying association between obesity and cancer risk can be categorized into the following headings:

(a) Adipokines, mainly leptin and adiponectin: adipokines activate intracellular pathways that are related to cell proliferation and survival such as JAK/STAT (Janus Kinase/Signal Transducer and Activator of Transcription), GSK3 (glycogen synthase kinase 3), MAPK (mitogen-activated protein kinase), PI3K/Akt (phosphatidylinositol 3-kinase/protein kinase B), AMPK (5′ AMP-activated protein kinase), and IRS (insulin receptor substrate). Moreover, the altered ratio of leptin and adiponectin has recently been widely studied for its potential role in tumor growth and metastasis.

(b) Insulin resistance: obesity is associated with increased circulating insulin and insulin-like growth factors, and these act as mitogens.

Table 5.3 List of adipokines

A1-acid glycoprotein
Acylation-stimulating protein (ASP)
Adiponectin
Adipose-specific fatty acid-binding protein (aP2)
Adipophilin
Adipsin
Agouti protein
Alpha macroglobulin
Angiopoietin-1, angiopoietin-2
Angiotensinogen
Apelin
Apolipoprotein E (apoE)
Autoxin
CC-chemokine ligand 2 (CCL2)
Ceruloplasmin
Chemerin
Cholesteryl ester transfer protein (CETP)
C-reactive protein (CRP)
CXC-chemokine ligand 5 (CXCL5)
Desnutrin
Fasting-induced adipose factor (FIAF)
Fibroblast growth factors
Fibronectin
Haptoglobin
Insulin-like growth factor 1 (IGF-1)
Interleukins (IL-1, IL-6, IL-8, IL-10, IL-17, IL-18)
Intercellular adhesion molecule-1 (ICAM-1)
Leptin
Lipocalin 2
Lipoprotein lipase
Macrophage migration inhibitory factor
Matrix metalloproteinase 1 (MMP1), MMP7, MMP9, MMP10, MMP11, MMP14, MMP15
Metallothionein
Monobutyrin
Monocyte chemoattractant protein-1 (MCP-1)
Nerve growth factor (NGF)
Omentin
Osteonectin
Pentraxin family member-3 (PTX-3)
Plasminogen activator inhibitor-1 (PAI-1)
Prostaglandins
Regulated upon activation, normally T cell expressed and secreted (RANTES)
Resistin and resistin-like molecules (RELM)
Retinol binding protein-4
Secreted frizzled-related protein 5 (SFRP5)
Serum amyloid A (SAA)

(continued)

Table 5.3 (continued)

Stromolysin
Tissue factor (TF)
Transforming growth factor beta
Tumor necrosis factor alpha
Type VI collagen
Vascular cell adhesion molecule-1 (VCAM-1)
Vascular endothelial growth factor (VEGF)
Vaspin
Visfatin
Zinc-a2-glycoprotein (ZAG)

(c) Inflammation: inflammatory cytokines have also been associated with cancer, and important molecules, such as interleukin-6 and tumor necrosis factor-α, are synthesized and secreted by adipocytes.

(d) Other underlying mechanisms: elevated circulating level of estrogen, caused by peripheral aromatization of androgens in adipose tissues due to increased cytochrome P450 activity, is also associated to estrogen-dependent cancers such as breast cancer; increased oxidative stress and peroxidation can increase the mutation rates; altered immune response can increase the survival of malignant cells; obesity-associated comorbidities such as hypertension, acid reflux, increase iodine uptake, and decreased vitamin D bioavailability can also predispose to certain cancer types; adipose tissue hypoxia (ATH) increases local insulin resistance and inflammation, and induces the synthesis of HIF-1α (hypoxia-inducible factor), associated with cancer [29].

These mechanisms are outlined in Fig. 5.1.

5.4.1 Adipokines

Adipokines are polypeptide hormones derived from adipocytes. Currently, there are more than 50 different types of known adipokines. In obesity, with the increase in adipose tissue mass and cell number, adipokines levels are altered. Some of the hypotheses linking adipokines and cancer are based on altered physiological states because of altered homeostatic levels of adipokines, while others attribute the adipokine's effects to their direct action on tumor cells. More than 50 types of adipokines have been described (Table 5.3), and leptin and adiponectin have been most widely studied for its possible role in outcome of obesity-related cancers.

Leptin

Leptin is a 16-kDa adipokine that is produced mainly by white adipose tissue. Others sites of production include the placenta, intestine, stomach, ovaries, bone marrow, brain, pituitary, liver, mammary epithelial cells, and skeletal muscle. Leptin

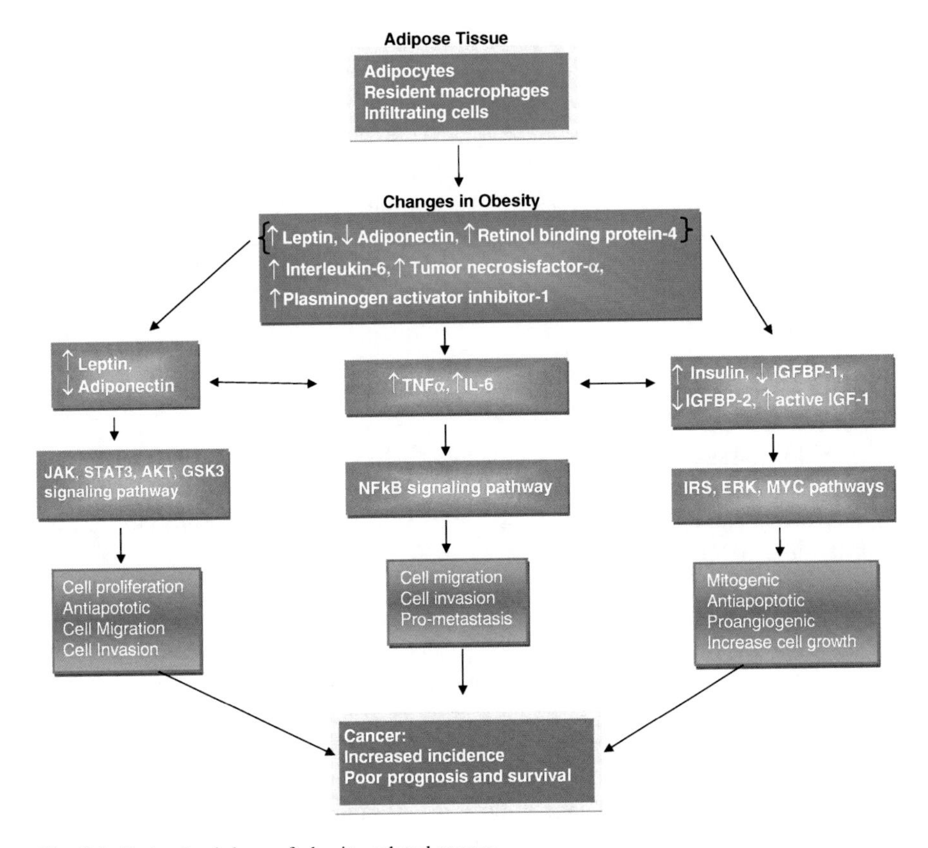

Fig. 5.1 Pathophysiology of obesity-related cancer

levels are positively correlated with the body mass. Its synthesis is influenced by insulin, tumor necrosis factor alpha (TNF-α), glucocorticoids, sex hormones, and prostaglandins. Its expression is also stimulated by hypoxia (commonly found in solid tumors), through the hypoxia-induced factor-1 (HIF-1) [30, 31].

The main role of leptin is to regulate energy homeostasis by balancing energy intake and energy expenditure, through its action on the arcuate nucleus of the hypothalamus. It has additional effects on the endocrine and immune systems including reproduction, glucose homeostasis, bone formation, tissue remodeling, and inflammation [32, 33]. Leptin binds to its receptor and activates different signaling pathways such as the JAK/STAT (Janus Kinase/ Signal Transducer and Activator of Transcription), MAPK (mitogen-activated protein kinase), PI3K/AKT (phosphatidylinositol 3-kinase/ protein-kinase B), AMPK (5′ AMP-activated protein kinase), and IRS (insulin receptor substrate) pathways, which affect cell proliferation and survival [34].

Congenital leptin deficiency is a rare condition associated with hyperphagia, impaired thermogenesis, immune defects, insulin resistance, dyslipidemia, hypogonadotropic hypogonadism, functional and structural alterations in the brain, and impairment of cognitive development, all reversible by leptin treatment [35–40].

Being the first discovered adipokine and highly increased in obesity, its role in the pathogenesis of cancer has been extensively investigated. Leptin is a pleiotropic hormone being mitogenic, antiapoptotic, proangiogenic, and proinflammatory in various cellular systems. Leptin stimulates growth, migration, and invasion in tumor cell models, which are relevant in the pathogenesis of cancer [30, 41]. Leptin also increases the production of cytokines by macrophages (such as IL-6, IL-12, and TNF-α), stimulating cancer cells [42]. Leptin appears to be involved in angiogenesis as well [43]. The role of leptin has been studied in different types of cancers and is described below:

Prostate Cancer

Epidemiological studies present contradictory evidences for the association between leptin and prostate cancer. A review summarizes those studies, with 3 out of 9 studies showing positive associations between blood leptin levels and prostate cancer [44]. In the Physicians' Health Study, a 25-year prospective study, no association between leptin level and prostate cancer was reported [45]. A genetic study demonstrated that the leptin polymorphism (-2548 G/A), leading to higher leptin levels, is associated with susceptibility to prostate cancer and risk of advanced disease [46]. It is suggested that, regardless of these contradictory data, leptin may be associated with more advanced, hormone-refractory prostate cancer [47].

Leptin receptors are present in prostate tissues and leptin induces growth and angiogenesis in prostate tissue [43, 48, 49]. In vitro, leptin stimulates growth of the androgen-insensitive prostate cancer lines DU145 and PC3 but not the androgen-sensitive cell lines, LNCaP [50, 51]. Leptin has been reported to promote the proliferation and survival of DU145 and PC3 cell lines through the activation of the PI3K and the classical MAPK pathways; additionally, leptin mediates the growth effect through JNK (c-Jun N-terminal kinase) MAP kinase pathway [50]. Leptin promotes neoangiogenesis in prostate cancer by inducing a number of proangiogenic factors including vascular endothelial growth factor (VEGF), fibroblast growth factor 2 (FGF 2), and the matrix metalloproteinases 2 and 9 [52].

Breast Cancer

High leptin levels contribute to the pathogenesis of breast cancer. Vona-Davies and Rose summarize the contradictory results found in case–control studies (one of which was nested within a prospective study), with 3/10 of results showing positive correlations between leptin levels and breast cancer [53]. Another prospective observational study demonstrated that BMI and leptin were significantly correlated with pathological tumor classifications and TNM stage in postmenopausal breast cancer patients [54]. It is hypothesized that leptin may increase breast cancer risk in postmenopausal women specifically [54, 55], in which the only source of estrogens is the adipose tissue.

The role of leptin in breast cancer can be explained on several levels (1) leptin induces the growth of breast cancer cells through the activation of the JAK/STAT3, MAPK-ERK1/2 (extracellular signal-regulated kinases 1/2), and/or PI3K pathways; (2) leptin mediates angiogenesis by inducing the expression of VEGF [43]; (3) leptin induces the transactivation of human epidermal growth factor receptor 2 (ErbB-2) and interacts with IGF-1 in triple negative breast cancer cells, transactivating the epidermal growth factor receptor (EGFR) and promoting invasion and migration; (4) leptin stimulates aromatase expression, increasing estrogen levels and affecting the growth of estrogen receptor (ER)-positive breast cancer cells; (5) leptin induces MAPK-dependent activation of ER [51, 56, 57]; and (6) leptin stimulates proteolytic cleavage of intercellular matrix, promoting cancer cell invasion [58].

In addition, leptin and its receptor are significantly overexpressed in human primary and metastatic breast cancer, being most abundant in less differentiated tumors [59]. In breast cancer cells, the overexpression of leptin is associated with the leptin promoter polymorphism Lep-2548G/A [60]. The subtype of leptin receptor seems to make a difference in the prognosis, since patients with elevated Ob-Ra expression have longer relapse-free survival as compared to patients with high Ob-Rb/Ob-Ra ratio [61]. The implication of leptin as a growth factor to breast cancer is further strengthened by the fact that leptin-deficient *ob/ob* [62] and leptin-resistant *db/db* mice [63] do not develop transgene-induced mammary tumors.

More recently, the role of leptin has been studied in the context of epithelia-to-mesenchymal transition and cancer stem cells. Leptin seems to trigger cancer stem cell differentiation in breast cancer, thereby promoting tumor cell survival and metastasis [64, 65].

Colorectal Cancer

Some prospective studies show positive correlation between leptin levels and colorectal cancer [66–68], while others show no association [69, 70]. Some studies observed significantly lower serum leptin concentration in patients with colorectal cancer, independent of BMI and weight loss [71, 72]. In less differentiated colorectal cancers, leptin expression is decreased [73], and the expression level may be positively correlated with a better prognosis [74].

Leptin is overexpressed in primary colorectal cancers, being significantly correlated with tumor grade and with the presence of adenocarcinoma [75]. Leptin promotes tumor growth via MAPK pathway (ERK 1/2 and JNK), JAK/STAT3, and PI3K/AKT and suppresses apoptosis [76–78]. The proangiogenic effect of leptin plays an additional role on the pathogenesis of colorectal cancers [43]. Leptin also induces IL-6 production by $Apc^{Min/+}$ colon epithelial cells, which leads to the growth and proliferation of preneoplastic cells [79]. However, in vivo studies are contradictory. In $Apc^{Min/+}$, *ob/ob* and *db/db* mice, leptin supplementation did not affect tumorigenesis [80, 81]. In a high-fat 1,2-dimethylhydrazine (a potent carcinogen)-treated rat model, growth of colonic cancer cells was enhanced by leptin [82], showing that leptin's effects may be synergistic to other environmental factors.

Thyroid Cancer

There is strong epidemiological evidence suggesting that obesity and increased BMI are positively correlated with thyroid cancer [83–85], but there are very few studies evaluating the association between serum leptin levels and thyroid cancer. In a small Turkish study, higher levels of serum leptin were present in patients with papillary thyroid carcinoma [86].

Leptin treatment may increase thyroid cell growth and function [87]. Leptin and leptin receptors are expressed in papillary thyroid cancer and are associated with aggressiveness [88, 89]. Thyroid cancer cell lines—anaplastic (ARO), follicular (WRO), and papillary (CGTH-W3)—express long-form leptin receptors, but leptin treatment does not alter the expression of the sodium-iodide symporter, cell growth, or cell cycle. However, it promotes cell migration of papillary thyroid cancer cells and inhibits migration of anaplastic and follicular cancer cells [90].

Renal Cancer

Epidemiological studies on the association between leptin levels and renal cancer are scarce. In a case–control study that included 70 patients with renal cell carcinoma, leptin was inversely associated with cancer risk (OR: 0.53, CI: 0.28–0.99, $p=0.05$), which the authors attribute to leptin's proimmunogenic effects [91]. Similarly, higher serum leptin was an independent predictor of progression-free survival [92].

Leptin receptor is present in Caki-1, ACHN, 769P, A498, SKRC44, and SKRC49 cells and in the murine renal cancer cell line Renca. Leptin induces invasiveness in murine cancer cell line model [93]. Leptin also increases the proliferation and mobility capabilities of Caki renal carcinoma cells by upregulating the expression of the JAK/STAT3 and ERK1/2 signaling pathways [94]. Some of the findings in vitro renal cancer models are contradictory to epidemiological evidences.

Endometrial Cancer

There are some evidences suggesting that leptin levels correlated with the presence of endometrial cancer, but after adjusting for BMI, no such correlation was found. It suggests that leptin may be not directly associated with endometrial cancer [95–97].

Leptin receptors are expressed by normal and malignant human endometrium and its expression is inhibited by treatment with progesterone [98]. In malignant tissues, the expression of leptin receptors is increased through the stimulatory effect of hypoxia-inducible factor 1α [99]. There are some reports of lower expression of the short form of the leptin receptor (Ob-Ra, which activates JAK2, IRS-1, and MAPK), suggesting that aberrant leptin receptor isoforms may be involved in the pathogenesis of endometrial cancer [96]. Leptin promotes endometrial cancer cell lines (ECC1 and Ishikawa endometrial adenocarcinoma cells) growth and invasiveness via MAPK and AKT pathways [100]. It enhances cyclin D1 expression (a cell cycle regulator required for completion of the G1/S transition) through JAK/STAT, MAPK, and PKA (protein kinase A) activation, leading to human endometrial cancer proliferation [101]. Additionally, leptin also increases several angiogenic factors [102].

Pancreatic Cancer

Epidemiological evidences suggested inverse correlation between leptin level and pancreatic cancer [103, 104], which may be just a consequences of weight loss observed in many cancer patients [105]. Also, in vitro, leptin inhibits growth of two human pancreatic cancer cell lines (PANC-1 and Mia-PaCa) [51]. In vivo studies in *ob/ob* and *db/db* mice showed that leptin deficiency or resistance is associated with larger tumors, increased number of metastases, and increased mortality. This suggests that other factors associated with obesity may play a role in the pathogenesis of pancreatic cancer but not leptin [106].

Adiponectin

Adiponectin is a 30-kDa peptide hormone, present physiologically as trimer, hexamer, or high-molecular-weight multimeric complexes [107]. It is expressed exclusively and in large quantities by adipocytes, making up for approximately 0.01 % of total human plasma protein content [108]; it is negatively regulated by several other products of adipose tissue that become elevated during obesity such as TNF-α, IL-6, and IL-18 [109]. Adiponectin levels are negatively correlated with body fat and BMI. The two isoforms of the adiponectin receptors, AdipoR1 and AdipoR2, are most abundantly found in striated muscle and the liver, respectively but are widely expressed in many types of normal and cancerous cells [110].

Despite high level of adiponectin in circulation, its functions are not clearly understood. Neither its deletion nor its overexpression leads to any significant body weight change in mice. Adiponectin acts as an insulin sensitizer [107]. It also acts as anti-inflammatory and antithermogenic agent. Additionally, it has been identified as a likely angiogenic inhibitor that maintains the balance of quiescent vasculature present in adult adipose tissue [111].

Nonetheless, adiponectin has been shown to play an active role in regulating neovascularization in tumors, suggesting that its effects may be more pronounced under pathological conditions, where there is more aggressive angiogenic activity than in slowly expanding adipose tissue [111].

Prostate Cancer

Adiponectin levels are inversely related with risk, incidence, and histological tumor grade of prostate cancer [112, 113]. It inhibits cell growth and proliferation in the metastatic prostate cancer cell lines DU145, PC3, and LNCaP by downregulation of STAT3 signaling. STAT3 increases the activity of androgen receptors, leading to an amplification of the mitogenic signal; this forms the basis of hormone-refractory prostate cancer and is crucial for the survival of DU145 cells [114]. It can also exert an inhibitory effect on cell proliferation, regardless of androgen dependence through an AMPK-mediated pathway. Binding of adiponectin to its receptors activates AMPK, which regulates a diverse array of metabolic functions within the cell; notably, it phosphorylates and activates TSC2, which is an inhibitor of mTOR. mTOR and its downstream effectors, for example the PI3K/AKT pathway, have been implicated in several cancers, including prostate cancer, in which the PTEN gene is frequently inactivated [112].

Breast Cancer

Epidemiological studies that suggest the association between adiponectin levels and breast cancer risk appears to be dependent on the menopausal state. In postmenopausal women with breast cancer, adiponectin levels are significantly decreased, while there is no such correlation in premenopausal women with breast cancer [115].

Adiponectin inhibits breast cancer cell proliferation in vitro, in part by inhibiting the TNF-α-mediated NF-κB signaling pathway. TNF-α is a proinflammatory cytokine, which promotes estrogen biosynthesis, especially in obese postmenopausal women, in whom adipose tissue is the main source of estrogens [116]. Elevated level of estrogen, stimulated by TNF-α, supports the survival and growth of breast cancer cells. It also suppresses VEGF expression by inhibiting TNF-α [112]. This lack of angiogenic support in a growing tumor can lead to considerable tumor cell apoptosis and reduction in tumor weights and volumes as demonstrated in adiponectin-treated mice by Brakenhielm et al. [111].

Adiponectin inhibits breast cancer cell invasion by activating the AMPK pathway [117]. In addition, it also stimulates the PPAR-α pathway in breast cancer, which has shown to increase nuclear levels of BRCA1, a tumor suppressor protein involved in DNA repair that reduces cancer risk [118].

The increased insulin resistance and hyperinsulinemia due to a fall in adiponectin in obesity also has an impact on breast cancer. This mechanism is described in a separate subheading.

Colorectal Cancer

There are contradictory epidemiological evidences for the association between colorectal cancer and adiponectin levels [69, 119], but many studies have shown that low circulating adiponectin is associated with an increased incidence of colorectal cancer [120–122], even after correcting for BMI and insulin resistance.

AdipoR1 and AdipoR2 expression is increased in colorectal cancer cells compared with normal tissues [123], giving support to the hypotheses that adiponectin may have some role in colorectal cancer. Some in vitro studies have shown that adiponectin may inhibit colorectal cancer cells through the activation of the AMPK pathway, which inhibits mTOR, leading to antiproliferative and proapoptotic effects of adiponectin [124]. Furthermore, adiponectin can reduce lipogenic activity—which is essential to maintain membrane integrity in the rapidly dividing tumor cells—by suppressing SREBP-1c activity. This in turn leads to a decrease in the expression of key lipogenic enzymes such as FAS [125].

The study by Fenton et al. highlights the suppression by adiponectin of two separate leptin-mediated pathways. Adiponectin phosphorylates and activates IκK, an inhibitor of NF-κB via MAPK pathway. Additionally, It also suppress the proinflammatory IL-6 trans-signaling induced by leptin, by decreasing the availability of the soluble IL-6 receptor (sIL-6R) and increasing sgp130, the natural inhibitor of the IL6–sIL-6R complex [126].

Thyroid Cancer

There is lack of studies evaluating the relationship between adiponectin and thyroid cancer. A follow-up study of patients with end-stage renal disease, who have a higher risk for developing cancer, showed that the most common site of cancer was the kidney (26.7 %), followed by thyroid (13.3 %) and stomach (13.3 %). Lower adiponectin was an independent predictor of malignancy, but the associations between hypoadiponectinemia and each type of cancer were not analyzed [127].

Renal Cancer

Studies have indicated strong inverse association between circulating levels of adiponectin and renal cancer [128–130]. Decreased levels of adiponectin are related with the presence of metastases, but its association with tumor grade appears to be inconsistent. The role of adiponectin in renal cancer is independent of BMI—in fact, Horiguchi et al. report a slight tendency for higher BMI to be associated with better clinicopathological features of the cancer, an observation that may be due to cancer-induced cachexia, characterized by a preferential loss of skeletal muscle rather than adipose tissue [130].

Both adiponectin receptors, especially AdipoR1, are expressed in normal renal tissue as well as renal cancer tumor cells, but receptors may be downregulated in cancer, reducing the protective role of adiponectin in tumor cells [128]. Low adiponectin levels are also implicated in the activation of several cancer-promoting factors including STAT3, ERK 1/2, and Akt.

Endometrial Cancer

The relationship between low circulating adiponectin levels and endometrial cancer has been established independently of BMI [131, 132], although a combination of a decrease in adiponectin levels with an increase in adipose tissue mass, as it often occurs, constitutes a greater risk [133]. Significantly, this relationship is especially strong in premenopausal women, contrary to that of breast cancer risk [115, 132, 133].

The action of adiponectin in endometrial cancer is very similar to that in breast cancer. Both tissues respond to estrogens, which promote cell proliferation; both tissues also express the proangiogenic factor VEGF. Adiponectin suppresses the biosynthesis of estrogens and the expression of VEGF through its effects on the NF-κB signaling pathway [132]. It also exerts its action by decreasing insulin resistance, as adiponectin acts as insulin sensitizer. The activation of AMPK is also an inhibiting effect of adiponectin on endometrial cancer [134].

A study by Cong et al. has expanded considerably the known mechanism of adiponectin through the AMPK pathway, by suggesting that adiponectin causes cell cycle arrest and induces apoptosis in endometrial carcinoma cell lines, HEC-1-A and RL95-2, via different pathways; the protein kinase Akt is inactivated and cyclin D1 is downregulated in HEC-1-A cells, whereas in RL95-2 cells there is a reduction in cyclin E2 expression, as well as an inhibition of p42/44 MAPK. These two cell lines differ genetically, as the RL95-2 line is deficient in the tumor suppressor PTEN, which regulates the PI3K signaling pathway [135].

Pancreatic Cancer

Epidemiological studies have suggested contradicting association between circulating level of adiponectin and pancreatic cancer: elevated adiponectin levels [136, 137], decreased [106, 138], and unchanged [103]. Notably, case–control studies [136, 137] supported the hypothesis that increased adiponectin is associated with pancreatic cancer, whereas one prospective study on male Finnish smokers reported an inverse correlation [138].

Adiponectin has been tied to cancer cachexia, and its increase may be due to weight loss caused by cancer progression [139]; this is in agreement with the observation that adiponectin levels are increased in anorexia nervosa or prolonged voluntary weight loss [140].

Despite a number of epidemiological studies, the direct action of adiponectin on pancreatic cancer cells has not yet been elucidated. It has been observed that NF-κB activation occurs in cell cultures, animal models, and human tissue of pancreatic cancer [141]. Similarly, the activation of carcinogenic pathways involving STAT3, mTOR, PI3K, and ERK, mediated by IL-6, have also been found to contribute to the survival and proliferation of pancreatic cancer cells [142].

Other Adipokines

Insulin-Like Growth Factor 1

Insulin-like growth factor 1 (IGF-1) is induced by growth hormone both locally in a paracrine/autocrine manner and in an endocrine manner via action on the liver. It is expressed in adipocytes and stromal cells (at levels comparable to those in the liver [143]), being important for the regulation of IGF-1 serum concentrations [144].

The actions of IGF-1 on cell proliferation, differentiation, migration, and survival signaling are crucial for tumor progression. It is one of the factors that enables cells to pass the G1-S checkpoint in the cell cycle [145]. There is substantial data from epidemiologic, animal, and in vitro studies indicating that the GH–IGF-1 axis has tremendous influence on cancer biology, cancer risk, and carcinogenesis [146].

Visfatin

Visfatin [pre-B-cell colony-enhancing factor (PBEF), nicotinamide mononucleotide denylyltransferase (NAMPT)] is a 52-kDa adipokine found in the cytoplasm and in the nucleus of cells of many tissues and organs including the brain, kidney, lung, spleen, and testis. It is preferentially expressed in visceral adipose tissue. Visfatin is released predominantly from macrophages rather than from adipocytes in visceral adipose tissues [147]. It has roles in vascular inflammation and angiogenesis and has been also implicated in the tumorigenesis and metastasis of various cancers such as colon, brain, pancreas, liver, stomach, prostate, and breast cancers [148]. Visfatin treatment increases the proliferation of human breast cancer cells and upregulates the expression of the matrix metalloproteinases MMP-2 and MMP-9, which play a major role in tumor cell metastasis [149].

Angiopoietin-Like 4

Angiopoietin-like 4 (ANGPTL4) is an adipokine also described as FIAF (fasting-induced adipose factor), PGAR [peroxisome proliferator-activated receptor (PPAR) γ-angiopoietin-related], and HFARP (hepatic fibrinogen/angiopoietin-related protein). It is expressed by the liver and white adipose tissue (adipocytes, monocytes, and endothelial cells), and it is involved in the regulation of lipid and glucose metabolism. More recently, it has been implicated in cancer growth and progression, anoikis resistance, altered redox regulation, angiogenesis, and metastasis [150]. ANGPTL4 is possibly an important redox player in cancer by maintaining an elevated oncogenic O_2:H_2O_2 ratio, and its inhibition may be potential therapeutic target [151].

Resistin

Resistin, also known as adipose tissue-specific secretory factor (ADSF) or C/EBP-epsilon-regulated myeloid-specific secreted cysteine-rich protein (XCP1), is secreted mainly by the adipose tissue and also by macrophages in humans. It is increased in the obese, and it is associated with increased insulin resistance (hence the name) via AMPK suppression [152]. In human prostate cancer cells, resistin has been shown to be overexpressed and to induce cell malignant cell proliferation through PI3K/Akt signaling pathways [153]. Moreover, the odds ratio of breast cancer for women in the highest quartile of serum resistin levels was 2.08 in a small study (95 % CI 1.04–3.85) [154]. Elevated levels of resistin have also been associated with different gastrointestinal cancer types [155].

Plasminogen Activator Inhibitor-1

Plasminogen Activator Inhibitor-1 (PAI-1) is produced by endothelial cells, stromal cells, and adipocytes mainly in visceral adipose tissue. It interacts with vitronectin, integrins, and other components of the uPA system and also affects the extracellular matrix, being involved in tumor growth, invasion, metastasis, and angiogenesis. It has also been suggested that PAI-1 inhibits apoptosis, regulates angiogenesis, and increases cell adhesion and migration. Elevated PAI-1 has been associated with poor prognosis for several types of cancer [156].

Apelin

Apelin is a peptide first identified in the gastrointestinal tract, but it is also expressed in various tissues including the heart, lung, liver, brain, and bone. In the adipose tissue, expression of apelin mRNA is higher on the stromal-vascular fraction than on the adipocytes [157]. Apelin was first identified as an endogenous ligand of the orphan G-protein-coupled receptor (GPCR) [158]. Apelin regulates blood pressure, fluid homeostasis, and cardiac contractility. It is an important factor in physiologic angiogenesis. Its overexpression increases tumor growth in a murine breast cancer model, increased apelin expression is associated with poor prognosis in lung cancer patients. In vivo, apelin stimulates tumor growth and tumor angiogenesis in various cancers including lung, breast, brain, and oral cancer [159].

5.4.2 *Insulin Resistance*

Obesity is the most critical factor in the emergence of metabolic diseases. Adipose tissue modulates metabolism by releasing nonessential fatty acids (NEFAs), glycerol, hormones (leptin and adiponectin), and proinflammatory cytokines [160]. There are several hypothesis proposed for insulin resistance in obesity. Retinol-binding protein-4 (RBP4) reduces PI3K signaling in muscle and enhances the expression of the phosphoenolpyruvate carboxykinase (a gluconeogenic enzyme) in liver, resulting in insulin resistance [161]. Moreover, insulin sensitizers, such as adiponectin, are decreased in obese individuals. Another mechanism describing insulin resistance in obesity is based on inflammation, induced by adipocyte-derived cytokines such as TNF-α, IL-6, and MCP-1. In addition, adipose tissue has resident macrophages and other cells that might also have roles on insulin resistance [162].

A number of epidemiological studies have independently linked insulin, C-peptide, IGFBP-1, IGFBP-2, and different types of cancer. High circulating levels of C-peptide are associated with the increased risk of postmenopausal breast cancer, colorectal cancer, pancreatic cancer, and endometrial cancer [163].

Most of the roles of IGFs in cancer have been studied in transgenic mice models. Transgenic mice with high expression of IGF-1 or IGF-1R have shown to develop skin cancer, prostate cancer, pancreatic neuroendocrine carcinoma, and breast cancer, while those expressing high levels of IGF-2 have developed lung cancer and breast cancer [164]. Similar studies promoted tumorigenesis by using inducible mice mammary tumor virus (MMTV) systems with transgenic overexpression of IGF-1R in mammary ductal epithelium [165]. The effect of IGF-1 on tumor development is more of a chronic outcome than acute. In one of the studies verifying this argument, genetic deletion of IGF-1 in the liver reduced liver metastases of colorectal carcinoma, while acute deletion did not have the same effect [166].

Prolonged hyperinsulinemia reduces the production of IGFBP-1 and IGFBP-2, leading to an increase in the active form of IGF-1. Increased active IGF-1 in the cellular environment favors tumor development by acting as a mitogen, promoting antiapoptotic activity, promoting angiogenesis, regulating cell size, increasing cell migration, and potentiating growth factors [25]. Elevated circulating levels of insulin, IGF-1 and IGF-2 activate insulin receptor substrate (IRS) proteins by phosphorylation, which in turn promotes cell proliferation via signal through growth factor receptor-bound protein 2 (GRB2) and guanine nucleotide exchange factor (SOS) to activate MAPK pathways. MAPK can also activate oncogenic pathways by phosphorylating transcriptional factors MYC and other members of ETS family, one of the largest families of transcription factors. Insulin resistance substrate (IRS) signaling cascade also activates PI3K, which phosphorylates AKT and activates mTOR. This increases protein translocation, increases cell size, and promotes cell growth. AKT inhibits cell death by inhibiting BCL-2 and promotes cell division by activating Cyclin D. Hence, these circulating peptides act as mitogens.

5.4.3 *Inflammation*

Obesity is associated with chronic inflammation. It is marked by abnormal cytokine production and activation of proinflammatory signaling. Also, adipose tissue contains resident macrophages, which are source of number of proinflammatory cytokines. Clinical and epidemiological studies have suggested a strong association between inflammation and different types of cancer. Inflammatory molecules can provide growth signals that promote the proliferation of malignant cells [167] and chronic inflammation increases cancer risk. Inflammation impacts every single step of tumorigenesis, from initiation through tumor promotion, all the way to metastatic progression [167, 168]. The two most important components of obesity-related inflammation, TNF-α and IL-6, are widely studied for their association with different types of cancer.

TNF-α, originally identified and named for its role in tumor necrosis, is widely described for its role in tumor promotion by inducing angiogenesis and increasing risk of metastasis development [164]. IL-6 is a proinflammatory cytokine, and circulating IL-6 correlates with BMI [169]. The adipose tissue accounts for 35 % of circulating IL-6 [170]. There are several studies suggesting the correlation between serum levels of IL-6 and TNF-α with patient survival in different types of cancer, namely prostate cancer [171], ovarian carcinoma [172], and pancreatic carcinoma [173]. TNF-α-308, a polymorphic form of TNF-α, has been confirmed as a risk factor for a range of cancer types such as breast cancer [174], gastric cancer [175], and hepatocellular cancer [176].

The role of TNF-α is established in mice models. In one study, TNF-α knockout animals developed delayed tumor compared with controls on treatment with tumor promoters such as okadaic acid and 12-*o*-tetradecanoylphorbol-12-acetate (TPA) [177]. Similarly, deletion of TNF receptor gene Tnfrsf1a abrogates the ability of a high-fat diet to promote liver carcinogenesis induced by diethyl-nitrosamine [178]. The two inflammatory pathways activated in adiposity are c-Jun NH2-terminal kinase (JNK) and IkB kinase-β (IKK-β) [25]. The nuclear factor-kB system is the most prominently described system, which correlates obesity-related inflammation with cancer. NF-kB is the downstream signal of IKK-B. NF-kB is activated by a variety of ligands including TNF-α, toll-like receptors, and other inflammatory cytokines.

IL-6 is the activator of signal transducer and activator of transcription 3 (STAT3). In genetic and dietary obese models of animals, STAT3 is activated. IL-6 knockout animal models are resistant to carcinogen-induced hepatocarcinoma [179]. Moreover, recent data suggest the existence of a relationship between cancer stem-like cells (CSCs) and interleukin-6 (IL-6) expression. The results of one study suggested that IL-6 may trigger a potential autocrine/paracrine activity of Notch-3 signaling by increased expression of Jagged-1 (transcriptional factor), which boosts the self-renewal breast CSCs. Likewise, Iliopoulos et al. revealed that NF-kB ensures high IL-6 levels both directly by activation of IL-6 transcription and indirectly by inhibition of let-7 microRNA [180].

Similarly, other proinflammatory factors derived from the adipose tissue factors, such as plasminogen activator inhibitor I (PAI-1), may play a role in the pathogenesis of cancer.

5.4.4 Other Mechanism Underlying Obesity and Cancer

The biosynthesis of estrogens differs significantly in premenopausal women and postmenopausal women. Estrogen is primarily synthesized in the ovary in premenopausal women, and in the periphery in postmenopausal women. In obese postmenopausal women, the adipose tissue is the main source of estrogen. Cytochrome P450 aromatase is the enzyme required in the biosynthesis of estrogen from androgens [181]. Aromatase is expressed in the normal breast adipose tissue as well as in tumor cells [182]. Estrogen levels are tenfold higher in breast tumors than in the circulation [183]. For more than 19 years, it has been established that serum estrogen levels can account for differences in breast cancer risk [184]. In a European prospective investigation into cancer and nutrition, serum androgens and estrogens in postmenopausal women were positively correlated with breast cancer risk, while serum levels of sex hormone binding globulin (SHBG) were inversely related with breast cancer risk [185]. In 1992, the National Cancer Institute conducted a study involving more than 13,000 healthy women considered at high risk for breast cancer. The group of women given tamoxifen (an estrogen receptor antagonist) had a lower rate of breast cancer than placebo [186].

Obesity is associated with several alterations in the levels of endogenous hormones. In pre- and postmenopausal women, excess weight and chronic hyperinsulinemia have been associated with changes in total and bioavailable plasma sex steroid levels. Weight-related increases in insulin and IGF-1 concentrations inhibit the hepatic synthesis of sex hormone binding globulin (SHBG) and enhance the synthesis of androgens by the gonads and adrenal glands. Furthermore, obese women have increased estrogen concentrations from peripheral conversion of androgens to estrogens in adipose tissue by aromatase enzyme. Consequently, pre- and postmenopausal obese women show a decrease in plasma SHBG levels, a rise in E1, and a rise in bioavailable testosterone unbound to SHBG. This leads to chronic anovulation in premenopausal women, with low progesterone levels. In these women, the lack of opposition by progesterone, added to the increase in IGF-1 and to the decrease in IGFBP-1, predisposes to the development of endometrial cancer. In postmenopausal women only, E2 is increased. Added to the increase in IGF-1, the risk for developing endometrial cancer is also increased in obese postmenopausal women, mainly due to estrogen excess [187].

Other factors associated with cancer in obese patients are (1) increased oxidative stress and peroxidation, which increase the mutation rates; (2) altered immune response, which increase the survival of mutated cells; (3) obesity-associated comorbidities such as hypertension, acid reflux, increase iodine uptake, and decreased vitamin D bioavailability, which can also predispose to certain cancer types; and (4) adipose tissue hypoxia (ATH), which increases local insulin resistance and chronic inflammation [188]. ATH also induces the synthesis of HIF-1α, which is associated with poor prognosis and increased metastases [188].

5.5 Recent Advances

Cancer cells are heterogeneous in nature and some subpopulations of cells have tumor-driving capacity. These cells have stem cell like properties and hence termed as cancer stem cells. Cancer stem cells are not only more invasive but are also drug resistant. It is suggested that leptin in involved in the induction of cancer stem cells [65]. Yan et al. described the role of leptin in epithelial-to-mesenchymal transition in breast cancer via β-catenin activation [64]. Epithelial-to-mesenchymal transition of cancer cell is considered the most critical step in metastases [189].

These reports are only scarce but there is growing evidence of investigations going in this field to elucidate the role of these adipokines in the outcome of cancer at molecular and cellular levels.

5.6 Conclusion

There are strong epidemiological, molecular, and clinical evidences showing associations between adipokines and the incidence and clinical outcome of cancer. While there are public health policies implemented to curb the rise of obesity rates, there is a parallel need to understand the role and mechanism of adipokines in the outcome of cancer. Future studies will lead to the development of prophylactic and therapeutic tools based on the pathophysiology of adipokines and obesity-related cancer.

References

1. World Health Organisation. Obesity and overweight. Fact Sheet No. 311. 2012.
2. Renehan AG, Roberts DL, Dive C. Obesity and cancer: pathophysiological and biological mechanisms. Arch Physiol Biochem. 2008;114(1):71–83.
3. Bergström A et al. Overweight as an avoidable cause of cancer in Europe. Int J Cancer. 2001;91(3):421–30.
4. IARC. Weight control and physical activity. IARC handbook of cancer prevention, vol. 6. Lyon: IARC Press; 2002.
5. Wiseman M. The Second World Cancer Research Fund/American Institute for Cancer Research expert report. Food, nutrition, physical activity, and the prevention of cancer: a global perspective. Proc Nutr Soc. 2008;67(03):253–6.
6. Reeves GK et al. Cancer incidence and mortality in relation to body mass index in the Million Women Study: cohort study. BMJ. 2007;335(7630):1134.
7. Harriss DJ et al. Lifestyle factors and colorectal cancer risk (1): systematic review and meta-analysis of associations with body mass index. Colorectal Dis. 2009;11(6):547–63.
8. Larsson SC, Wolk A. Obesity and colon and rectal cancer risk: a meta-analysis of prospective studies. Am J Clin Nutr. 2007;86(3):556–65.
9. Moghaddam AA, Woodward M, Huxley R. Obesity and risk of colorectal cancer: a meta-analysis of 31 studies with 70,000 events. Cancer Epidemiol Biomarkers Prev. 2007;16(12):2533–47.
10. Dai Z, Xu YC, Niu L. Obesity and colorectal cancer risk: a meta-analysis of cohort studies. World J Gastroenterol. 2007;13(31):4199–206.
11. Larsson SC, Wolk A. Obesity and the risk of gallbladder cancer: a meta-analysis. Br J Cancer. 2007;96(9):1457–61.
12. Larsson SC, Wolk A. Overweight, obesity and risk of liver cancer: a meta-analysis of cohort studies. Br J Cancer. 2007;97(7):1005–8.
13. Saunders D et al. Systematic review: the association between obesity and hepatocellular carcinoma—epidemiological evidence. Aliment Pharmacol Ther. 2010;31(10):1051–63.
14. Mathew A, George PS, Ildaphonse G. Obesity and kidney cancer risk in women: a meta-analysis (1992–2008). Asian Pac J Cancer Prev. 2009;10(3):471–8.

15. Ildaphonse G, George PS, Mathew A. Obesity and kidney cancer risk in men: a meta-analysis (1992–2008). Asian Pac J Cancer Prev. 2009;10(2):279–86.
16. Stolzenberg-Solomon RZ et al. Adiposity, physical activity, and pancreatic cancer in the National Institutes of Health–AARP Diet and Health Cohort. Am J Epidemiol. 2008;167(5): 586–97.
17. Arslan AA et al. Anthropometric measures, body mass index, and pancreatic cancer: a pooled analysis from the Pancreatic Cancer Cohort Consortium (PanScan). Arch Intern Med. 2010;170(9):791–802.
18. Olsen CM et al. Obesity and the risk of epithelial ovarian cancer: a systematic review and meta-analysis. Eur J Cancer. 2007;43(4):690–709.
19. Purdie DM et al. Body size and ovarian cancer: case–control study and systematic review (Australia). Cancer Causes Control. 2001;12(9):855–63.
20. Suzuki R et al. Body weight and incidence of breast cancer defined by estrogen and progesterone receptor status—a meta-analysis. Int J Cancer. 2009;124(3):698–712.
21. MacInnis R, English D. Body size and composition and prostate cancer risk: systematic review and meta-regression analysis. Cancer Causes Control. 2006;17(8):989–1003.
22. Larsson SC, Wolk A. Body mass index and risk of multiple myeloma: a meta-analysis. Int J Cancer. 2007;121(11):2512–6.
23. Larsson SC, Wolk A. Obesity and risk of non-Hodgkin's lymphoma: a meta-analysis. Int J Cancer. 2007;121(7):1564–70.
24. Calle EE et al. Overweight, obesity, and mortality from cancer in a prospectively studied cohort of U.S. adults. N Engl J Med. 2003;348(17):1625–38.
25. Roberts DL, Dive C, Renehan AG. Biological mechanisms linking obesity and cancer risk: new perspectives. Annu Rev Med. 2010;61(1):301–16.
26. Sjöström L et al. Effects of bariatric surgery on cancer incidence in obese patients in Sweden (Swedish Obese Subjects Study): a prospective, controlled intervention trial. Lancet Oncol. 2009;10(7):653–62.
27. Melmed S, Polonsky KS, Larsen PR, Kronberg HM. Williams text book of endocrinology. 12th ed. Orlando, FL: W.B. Saunders; 2011.
28. Maeda K, Okubo K, Shimomura I, Mizuno K, Matsuzawa Y, Matsubara K. Analysis of an expression profile of genes in the human adipose tissue. Gene. 1997;190:227–35.
29. Calzada MJ, del Peso L. Hypoxia-inducible factors and cancer. Clin Transl Oncol. 2007;9(5):278–89.
30. Garofalo C, Surmacz E. Leptin and cancer. J Cell Physiol. 2006;207(1):12–22.
31. Paz-Filho G et al. Associations between adipokines and obesity-related cancer. Front Biosci. 2011;16:1634–50.
32. Kelesidis T et al. Narrative review: the role of leptin in human physiology: emerging clinical applications. Ann Intern Med. 2010;152(2):93–100.
33. Boguszewski CL, Paz-Filho G, Velloso LA. Neuroendocrine body weight regulation: integration between fat tissue, gastrointestinal tract, and the brain. Endokrynol Pol. 2010; 61(2):194–206.
34. Fruhbeck G. Intracellular signalling pathways activated by leptin. Biochem J. 2006;393(Pt 1):7–20.
35. Baicy K et al. Leptin replacement alters brain response to food cues in genetically leptin-deficient adults. Proc Natl Acad Sci USA. 2007;104(46):18276–9.
36. Farooqi IS et al. Beneficial effects of leptin on obesity, T cell hyporesponsiveness, and neuroendocrine/metabolic dysfunction of human congenital leptin deficiency. J Clin Invest. 2002;110(8):1093–103.
37. Licinio J et al. Phenotypic effects of leptin replacement on morbid obesity, diabetes mellitus, hypogonadism, and behavior in leptin-deficient adults. Proc Natl Acad Sci USA. 2004; 101(13):4531–6.
38. Matochik JA et al. Effect of leptin replacement on brain structure in genetically leptin-deficient adults. J Clin Endocrinol Metab. 2005;90(5):2851–4.

39. Paz-Filho G et al. Changes in insulin sensitivity during leptin replacement therapy in leptin-deficient patients. Am J Physiol Endocrinol Metab. 2008;295(6):E1401–8.
40. Paz-Filho GJ et al. Leptin replacement improves cognitive development. PLoS One. 2008;3(8):e3098.
41. Bouloumie A et al. Leptin, the product of Ob gene, promotes angiogenesis. Circ Res. 1998;83(10):1059–66.
42. Trayhurn P, Wood IS. Adipokines: inflammation and the pleiotropic role of white adipose tissue. Br J Nutr. 2004;92(3):347–55.
43. Sierra-Honigmann MR et al. Biological action of leptin as an angiogenic factor. Science. 1998;281(5383):1683–6.
44. Hsing AW, Sakoda LC, Chua Jr S. Obesity, metabolic syndrome, and prostate cancer. Am J Clin Nutr. 2007;86(3):s843–57.
45. Li H et al. A 25-year prospective study of plasma adiponectin and leptin concentrations and prostate cancer risk and survival. Clin Chem. 2010;56(1):34–43.
46. Ribeiro R et al. Overexpressing leptin genetic polymorphism (−2548 G/A) is associated with susceptibility to prostate cancer and risk of advanced disease. Prostate. 2004;59(3):268–74.
47. Freedland SJ, Platz EA. Obesity and prostate cancer: making sense out of apparently conflicting data. Epidemiol Rev. 2007;29:88–97.
48. Frankenberry KA et al. Leptin induces cell migration and the expression of growth factors in human prostate cancer cells. Am J Surg. 2004;188(5):560–5.
49. Somasundar P et al. Prostate cancer cell proliferation is influenced by leptin. J Surg Res. 2004;118(1):71–82.
50. Onuma M et al. Prostate cancer cell-adipocyte interaction: leptin mediates androgen-independent prostate cancer cell proliferation through c-Jun NH2-terminal kinase. J Biol Chem. 2003;278(43):42660–7.
51. Somasundar P et al. Differential effects of leptin on cancer in vitro. J Surg Res. 2003;113(1):50–5.
52. Park HY et al. Potential role of leptin in angiogenesis: leptin induces endothelial cell proliferation and expression of matrix metalloproteinases in vivo and in vitro. Exp Mol Med. 2001;33(2):95–102.
53. Vona-Davis L, Rose DP. Adipokines as endocrine, paracrine, and autocrine factors in breast cancer risk and progression. Endocr Relat Cancer. 2007;14(2):189–206.
54. Maccio A et al. Correlation of body mass index and leptin with tumor size and stage of disease in hormone-dependent postmenopausal breast cancer: preliminary results and therapeutic implications. J Mol Med. 2010;88(7):677–86.
55. Rose DP, Komninou D, Stephenson GD. Obesity, adipocytokines, and insulin resistance in breast cancer. Obes Rev. 2004;5(3):153–65.
56. Cirillo D et al. Leptin signaling in breast cancer: an overview. J Cell Biochem. 2008;105(4):956–64.
57. Brown KA, Simpson ER. Obesity and breast cancer: progress to understanding the relationship. Cancer Res. 2010;70(1):4–7.
58. Castellucci M et al. Leptin modulates extracellular matrix molecules and metalloproteinases: possible implications for trophoblast invasion. Mol Hum Reprod. 2000;6(10):951–8.
59. Garofalo C et al. Increased expression of leptin and the leptin receptor as a marker of breast cancer progression: possible role of obesity-related stimuli. Clin Cancer Res. 2006;12(5):1447–53.
60. Terrasi M et al. Functional analysis of the -2548G/A leptin gene polymorphism in breast cancer cells. Int J Cancer. 2009;125(5):1038–44.
61. Revillion F et al. Messenger RNA expression of leptin and leptin receptors and their prognostic value in 322 human primary breast cancers. Clin Cancer Res. 2006;12(7 Pt 1):2088–94.
62. Cleary MP et al. Genetically obese MMTV-TGF-alpha/Lep(ob)Lep(ob) female mice do not develop mammary tumors. Breast Cancer Res Treat. 2003;77(3):205–15.
63. Cleary MP et al. Leptin receptor-deficient MMTV-TGF-alpha/Lepr(db)Lepr(db) female mice do not develop oncogene-induced mammary tumors. Exp Biol Med (Maywood). 2004;229(2):182–93.

64. Yan D et al. Leptin-induced epithelial-mesenchymal transition in breast cancer cells requires β-catenin activation via Akt/GSK3- and MTA1/Wnt1 protein-dependent pathways. J Biol Chem. 2012;287(11):8598–612.
65. Park J, Scherer PE. Leptin and cancer: from cancer stem cells to metastasis. Endocr Relat Cancer. 2011;18(4):C25–9.
66. Stattin P et al. Obesity and colon cancer: does leptin provide a link? Int J Cancer. 2004; 109(1):149–52.
67. Stattin P et al. Plasma leptin and colorectal cancer risk: a prospective study in Northern Sweden. Oncol Rep. 2003;10(6):2015–21.
68. Tamakoshi K et al. Leptin is associated with an increased female colorectal cancer risk: a nested case–control study in Japan. Oncology. 2005;68(4–6):454–61.
69. Nakajima TE et al. Adipocytokines as new promising markers of colorectal tumors: adiponectin for colorectal adenoma, and resistin and visfatin for colorectal cancer. Cancer Sci. 2010;101(5):1286–91.
70. Tessitore L et al. Leptin expression in colorectal and breast cancer patients. Int J Mol Med. 2000;5(4):421–6.
71. Kumor A et al. Serum leptin, adiponectin, and resistin concentration in colorectal adenoma and carcinoma (CC) patients. Int J Colorectal Dis. 2009;24(3):275–81.
72. Bolukbas FF et al. Serum leptin concentration and advanced gastrointestinal cancers: a case controlled study. BMC Cancer. 2004;4:29.
73. Koda M et al. Expression of the obesity hormone leptin and its receptor correlates with hypoxia-inducible factor-1 alpha in human colorectal cancer. Ann Oncol. 2007;18 Suppl 6:vi116–9.
74. Paik SS et al. Leptin expression correlates with favorable clinicopathologic phenotype and better prognosis in colorectal adenocarcinoma. Ann Surg Oncol. 2009;16(2):297–303.
75. Koda M et al. Overexpression of the obesity hormone leptin in human colorectal cancer. J Clin Pathol. 2007;60(8):902–6.
76. Hardwick JC et al. Leptin is a growth factor for colonic epithelial cells. Gastroenterology. 2001;121(1):79–90.
77. Uddin S et al. Leptin receptor expression in Middle Eastern colorectal cancer and its potential clinical implication. Carcinogenesis. 2009;30(11):1832–40.
78. Ogunwobi OO, Beales IL. The anti-apoptotic and growth stimulatory actions of leptin in human colon cancer cells involves activation of JNK mitogen activated protein kinase, JAK2 and PI3 kinase/Akt. Int J Colorectal Dis. 2007;22(4):401–9.
79. Fenton JI et al. Leptin, insulin-like growth factor-1, and insulin-like growth factor-2 are mitogens in ApcMin/+ but not Apc+/+ colonic epithelial cell lines. Cancer Epidemiol Biomarkers Prev. 2005;14(7):1646–52.
80. Aparicio T et al. Leptin stimulates the proliferation of human colon cancer cells in vitro but does not promote the growth of colon cancer xenografts in nude mice or intestinal tumorigenesis in Apc(Min/+) mice. Gut. 2005;54(8):1136–45.
81. Ealey KN, Lu S, Archer MC. Development of aberrant crypt foci in the colons of ob/ob and db/db mice: evidence that leptin is not a promoter. Mol Carcinog. 2008;47(9):667–77.
82. Liu Z et al. High fat diet enhances colonic cell proliferation and carcinogenesis in rats by elevating serum leptin. Int J Oncol. 2001;19(5):1009–14.
83. Engeland A et al. Body size and thyroid cancer in two million Norwegian men and women. Br J Cancer. 2006;95(3):366–70.
84. Clavel-Chapelon F et al. Risk of differentiated thyroid cancer in relation to adult weight, height and body shape over life: the French E3N cohort. Int J Cancer. 2010;126(12): 2984–90.
85. Leitzmann MF et al. Prospective study of body mass index, physical activity and thyroid cancer. Int J Cancer. 2010;126(12):2947–56.
86. Akinci M et al. Leptin levels in thyroid cancer. Asian J Surg. 2009;32(4):216–23.
87. Nowak KW et al. Rat thyroid gland expresses the long form of leptin receptors, and leptin stimulates the function of the gland in euthyroid non-fasted animals. Int J Mol Med. 2002; 9(1):31–4.

88. Cheng SP et al. Clinicopathologic significance of leptin and leptin receptor expressions in papillary thyroid carcinoma. Surgery. 2010;147(6):847–53.
89. Uddin S et al. Leptin-R and its association with PI3K/AKT signaling pathway in papillary thyroid carcinoma. Endocr Relat Cancer. 2010;17(1):191–202.
90. Cheng SP et al. Differential roles of leptin in regulating cell migration in thyroid cancer cells. Oncol Rep. 2010;23(6):1721–7.
91. Spyridopoulos TN et al. Inverse association of leptin levels with renal cell carcinoma: results from a case–control study. Hormones (Athens). 2009;8(1):39–46.
92. Horiguchi A et al. Increased serum leptin levels and over expression of leptin receptors are associated with the invasion and progression of renal cell carcinoma. J Urol. 2006;176(4 Pt 1):1631–5.
93. Horiguchi A et al. Leptin promotes invasiveness of murine renal cancer cells via extracellular signal-regulated kinases and rho dependent pathway. J Urol. 2006;176(4 Pt 1):1636–41.
94. Li L et al. Concomitant activation of the JAK/STAT3 and ERK1/2 signaling is involved in leptin-mediated proliferation of renal cell carcinoma Caki-2 cells. Cancer Biol Ther. 2008;7(11):1787–92.
95. Petridou E et al. Leptin and body mass index in relation to endometrial cancer risk. Ann Nutr Metab. 2002;46(3–4):147–51.
96. Yuan SS et al. Aberrant expression and possible involvement of the leptin receptor in endometrial cancer. Gynecol Oncol. 2004;92(3):769–75.
97. Cymbaluk A, Chudecka-Glaz A, Rzepka-Gorska I. Leptin levels in serum depending on body mass index in patients with endometrial hyperplasia and cancer. Eur J Obstet Gynecol Reprod Biol. 2008;136(1):74–7.
98. Koshiba H et al. Progesterone inhibition of functional leptin receptor mRNA expression in human endometrium. Mol Hum Reprod. 2001;7(6):567–72.
99. Koda M et al. Expression of leptin, leptin receptor, and hypoxia-inducible factor 1 alpha in human endometrial cancer. Ann N Y Acad Sci. 2007;1095:90–8.
100. Sharma D et al. Leptin promotes the proliferative response and invasiveness in human endometrial cancer cells by activating multiple signal-transduction pathways. Endocr Relat Cancer. 2006;13(2):629–40.
101. Catalano S et al. Evidence that leptin through STAT and CREB signaling enhances cyclin D1 expression and promotes human endometrial cancer proliferation. J Cell Physiol. 2009;218(3):490–500.
102. Carino C et al. Leptin regulation of proangiogenic molecules in benign and cancerous endometrial cells. Int J Cancer. 2008;123(12):2782–90.
103. Pezzilli R et al. Serum leptin, but not adiponectin and receptor for advanced glycation end products, is able to distinguish autoimmune pancreatitis from both chronic pancreatitis and pancreatic neoplasms. Scand J Gastroenterol. 2010;45(1):93–9.
104. Dalamaga M et al. Low circulating adiponectin and resistin, but not leptin, levels are associated with multiple myeloma risk: a case–control study. Cancer Causes Control. 2009;20(2): 193–9.
105. Brown DR, Berkowitz DE, Breslow MJ. Weight loss is not associated with hyperleptinemia in humans with pancreatic cancer. J Clin Endocrinol Metab. 2001;86(1):162–6.
106. Zyromski NJ et al. Obesity potentiates the growth and dissemination of pancreatic cancer. Surgery. 2009;146(2):258–63.
107. Ahima RS. Adipose tissue as an endocrine organ. Obesity (Silver Spring). 2006;14 Suppl 5:242S–9.
108. Wozniak SE et al. Adipose tissue: the new endocrine organ? A review article. Dig Dis Sci. 2009;54(9):1847–56.
109. Liu M, Liu F. Transcriptional and post-translational regulation of adiponectin. Biochem J. 2009;425(1):41–52.
110. Kadowaki T, Yamauchi T. Adiponectin and adiponectin receptors. Endocr Rev. 2005;26(3): 439–51.

111. Brakenhielm E et al. Adiponectin-induced antiangiogenesis and antitumor activity involve caspase-mediated endothelial cell apoptosis. Proc Natl Acad Sci USA. 2004;101(8):2476–81.
112. Kelesidis I, Kelesidis T, Mantzoros CS. Adiponectin and cancer: a systematic review. Br J Cancer. 2006;94(9):1221–5.
113. Goktas S et al. Prostate cancer and adiponectin. Urology. 2005;65(6):1168–72.
114. Bub JD, Miyazaki T, Iwamoto Y. Adiponectin as a growth inhibitor in prostate cancer cells. Biochem Biophys Res Commun. 2006;340(4):1158–66.
115. Mantzoros C et al. Adiponectin and breast cancer risk. J Clin Endocrinol Metab. 2004;89(3): 1102–7.
116. Cleary MP, Grossmann ME. Minireview: obesity and breast cancer: the estrogen connection. Endocrinology. 2009;150(6):2537–42.
117. Kim KY et al. Adiponectin-activated AMPK stimulates dephosphorylation of AKT through protein phosphatase 2A activation. Cancer Res. 2009;69(9):4018–26.
118. Lorincz AM, Sukumar S. Molecular links between obesity and breast cancer. Endocr Relat Cancer. 2006;13(2):279–92.
119. Lukanova A et al. Serum adiponectin is not associated with risk of colorectal cancer. Cancer Epidemiol Biomarkers Prev. 2006;15(2):401–2.
120. Wei EK et al. Low plasma adiponectin levels and risk of colorectal cancer in men: a prospective study. J Natl Cancer Inst. 2005;97(22):1688–94.
121. Otake S et al. Association of visceral fat accumulation and plasma adiponectin with colorectal adenoma: evidence for participation of insulin resistance. Clin Cancer Res. 2005;11(10): 3642–6.
122. Yamaji T et al. Interaction between adiponectin and leptin influences the risk of colorectal adenoma. Cancer Res. 2010;70(13):5430–7.
123. Barb D et al. Adiponectin in relation to malignancies: a review of existing basic research and clinical evidence. Am J Clin Nutr. 2007;86(3):s858–66.
124. Sugiyama M et al. Adiponectin inhibits colorectal cancer cell growth through the AMPK/ mTOR pathway. Int J Oncol. 2009;34(2):339–44.
125. Kim AY et al. Adiponectin represses colon cancer cell proliferation via AdipoR1- and -R2-mediated AMPK activation. Mol Endocrinol. 2010;24(7):1441–52.
126. Fenton JI et al. Adiponectin blocks multiple signaling cascades associated with leptin-induced cell proliferation in Apc Min/+ colon epithelial cells. Int J Cancer. 2008;122(11): 2437–45.
127. Park JT et al. Insulin resistance and lower plasma adiponectin increase malignancy risk in nondiabetic continuous ambulatory peritoneal dialysis patients. Metabolism. 2011;60(1): 121–6.
128. Pinthus JH et al. Lower plasma adiponectin levels are associated with larger tumor size and metastasis in clear-cell carcinoma of the kidney. Eur Urol. 2008;54(4):866–73.
129. Spyridopoulos TN et al. Low adiponectin levels are associated with renal cell carcinoma: a case–control study. Int J Cancer. 2007;120(7):1573–8.
130. Horiguchi A et al. Decreased serum adiponectin levels in patients with metastatic renal cell carcinoma. Jpn J Clin Oncol. 2008;38(2):106–11.
131. Soliman PT et al. Association between adiponectin, insulin resistance, and endometrial cancer. Cancer. 2006;106(11):2376–81.
132. Cust AE et al. Plasma adiponectin levels and endometrial cancer risk in pre- and postmenopausal women. J Clin Endocrinol Metab. 2007;92(1):255–63.
133. Dal Maso L et al. Circulating adiponectin and endometrial cancer risk. J Clin Endocrinol Metab. 2004;89(3):1160–3.
134. Takemura Y et al. Expression of adiponectin receptors and its possible implication in the human endometrium. Endocrinology. 2006;147(7):3203–10.
135. Cong L et al. Human adiponectin inhibits cell growth and induces apoptosis in human endometrial carcinoma cells, HEC-1-A and RL95 2. Endocr Relat Cancer. 2007;14(3):713–20.
136. Chang MC et al. Adiponectin as a potential differential marker to distinguish pancreatic cancer and chronic pancreatitis. Pancreas. 2007;35(1):16–21.

137. Dalamaga M et al. Pancreatic cancer expresses adiponectin receptors and is associated with hypoleptinemia and hyperadiponectinemia: a case–control study. Cancer Causes Control. 2009;20(5):625–33.
138. Stolzenberg-Solomon RZ et al. Prediagnostic adiponectin concentrations and pancreatic cancer risk in male smokers. Am J Epidemiol. 2008;168(9):1047–55.
139. Wolf I et al. Adiponectin, ghrelin, and leptin in cancer cachexia in breast and colon cancer patients. Cancer. 2006;106(4):966–73.
140. Brichard SM, Delporte ML, Lambert M. Adipocytokines in anorexia nervosa: a review focusing on leptin and adiponectin. Horm Metab Res. 2003;35(6):337–42.
141. Garcea G et al. Role of inflammation in pancreatic carcinogenesis and the implications for future therapy. Pancreatology. 2005;5(6):514–29.
142. Chen J, Huang XF. Interleukin-6 promotes carcinogenesis through multiple signal pathways. Comment on: clinical significance of interleukin-6 gene polymorphism and IL-6 serum level in pancreatic adenocarcinoma and chronic pancreatitis. Dig Dis Sci. 2009;54(6):1373–4.
143. Villafuerte BC et al. Expressions of leptin and insulin-like growth factor-I are highly correlated and region-specific in adipose tissue of growing rats. Obes Res. 2000;8(9):646–55.
144. Kloting N et al. Autocrine IGF-1 action in adipocytes controls systemic IGF-1 concentrations and growth. Diabetes. 2008;57(8):2074–82.
145. Stull MA et al. Requirement for IGF-I in epidermal growth factor-mediated cell cycle progression of mammary epithelial cells. Endocrinology. 2002;143(5):1872–9.
146. Renehan AG et al. Insulin-like growth factor (IGF)-I, IGF binding protein-3, and cancer risk: systematic review and meta-regression analysis. Lancet. 2004;363(9418):1346–53.
147. Curat CA et al. Macrophages in human visceral adipose tissue: increased accumulation in obesity and a source of resistin and visfatin. Diabetologia. 2006;49(4):744–7.
148. Bi TQ, Che XM. Nampt/PBEF/visfatin and cancer. Cancer Biol Ther. 2010;10(2):119–25.
149. Kim JG et al. Visfatin stimulates proliferation of MCF-7 human breast cancer cells. Mol Cells. 2010;30(4):341–5.
150. Tan MJ et al. Emerging roles of angiopoietin-like 4 in human cancer. Mol Cancer Res. 2012;10(6):677–88.
151. Zhu P et al. Angiopoietin-like 4 protein elevates the prosurvival intracellular O2(−):H2O2 ratio and confers anoikis resistance to tumors. Cancer Cell. 2011;19(3):401–15.
152. Qi Y et al. Loss of resistin improves glucose homeostasis in leptin deficiency. Diabetes. 2006;55(11):3083–90.
153. Kim HJ et al. Expression of resistin in the prostate and its stimulatory effect on prostate cancer cell proliferation. BJU Int. 2011;108(2 Pt 2):E77–83.
154. Sun CA et al. Adipocytokine resistin and breast cancer risk. Breast Cancer Res Treat. 2010;123(3):869–76.
155. Tiaka EK et al. The implication of adiponectin and resistin in gastrointestinal diseases. Cytokine Growth Factor Rev. 2011;22(2):109–19.
156. Prieto-Hontoria PL et al. Role of obesity-associated dysfunctional adipose tissue in cancer: a molecular nutrition approach. Biochim Biophys Acta. 2011;1807(6):664–78.
157. Garcia-Diaz D et al. Adiposity dependent apelin gene expression: relationships with oxidative and inflammation markers. Mol Cell Biochem. 2007;305(1–2):87–94.
158. Tatemoto K et al. Isolation and characterization of a novel endogenous peptide ligand for the human APJ receptor. Biochem Biophys Res Commun. 1998;251(2):471–6.
159. Heo K et al. Hypoxia-induced up-regulation of apelin is associated with a poor prognosis in oral squamous cell carcinoma patients. Oral Oncol. 2012;48(6):500–6.
160. Kahn SE, Hull RL, Utzschneider KM. Mechanisms linking obesity to insulin resistance and type 2 diabetes. Nature. 2006;444(7121):840–6.
161. Yang Q et al. Serum retinol binding protein 4 contributes to insulin resistance in obesity and type 2 diabetes. Nature. 2005;436(7049):356–62.
162. Shulman GI. Cellular mechanisms of insulin resistance. J Clin Invest. 2000;106(2):171–6.
163. Pisani P. Hyper-insulinaemia and cancer, meta-analyses of epidemiological studies. Arch Physiol Biochem. 2008;114(1):63–70.

164. Khandekar MJ, Cohen P, Spiegelman BM. Molecular mechanisms of cancer development in obesity. Nat Rev Cancer. 2011;11(12):886–95.
165. Jones RA et al. Transgenic overexpression of IGF-IR disrupts mammary ductal morphogenesis and induces tumor formation. Oncogene. 2006;26(11):1636–44.
166. Wu Y et al. Insulin-like growth factor-I regulates the liver microenvironment in obese mice and promotes liver metastasis. Cancer Res. 2010;70(1):57–67.
167. Grivennikov SI, Greten FR, Karin M. Immunity, inflammation, and cancer. Cell. 2010; 140(6):883–99.
168. de Visser KE, Eichten A, Coussens LM. Paradoxical roles of the immune system during cancer development. Nat Rev Cancer. 2006;6(1):24–37.
169. Kern PA et al. Adipose tissue tumor necrosis factor and interleukin-6 expression in human obesity and insulin resistance. Am J Physiol Endocrinol Metab. 2001;280(5):E745–51.
170. Mohamed-Ali V et al. Subcutaneous adipose tissue releases interleukin-6, but not tumor necrosis factor-α, in vivo. J Clin Endocrinol Metab. 1997;82(12):4196–200.
171. Michalaki V et al. Serum levels of IL-6 and TNF-[alpha] correlate with clinicopathological features and patient survival in patients with prostate cancer. Br J Cancer. 2004;90(12): 2312–6.
172. Charles KA et al. The tumor-promoting actions of TNF-α involve TNFR1 and IL-17 in ovarian cancer in mice and humans. J Clin Invest. 2009;119(10):3011–23.
173. Karayiannakis AJ et al. Serum levels of tumor necrosis factor-alpha and nutritional status in pancreatic cancer patients. Anticancer Res. 2001;21(2B):1355–8.
174. Shen C et al. Polymorphisms of tumor necrosis factor-alpha and breast cancer risk: a meta-analysis. Breast Cancer Res Treat. 2011;126(3):763–70.
175. Lu PH et al. Meta-analysis of association of tumor necrosis factor alpha-308 gene promoter polymorphism with gastric cancer. Zhonghua Yu Fang Yi Xue Za Zhi. 2010;44(3):209–14.
176. Yang Y et al. The TNF-α, IL-1B and IL-10 polymorphisms and risk for hepatocellular carcinoma: a meta-analysis. J Cancer Res Clin Oncol. 2011;137(6):947–52.
177. Suganuma M et al. Essential role of tumor necrosis factor α (TNF-α) in tumor promotion as revealed by TNF-α-deficient mice. Cancer Res. 1999;59(18):4516–8.
178. Park EJ et al. Dietary and genetic obesity promote liver inflammation and tumorigenesis by enhancing IL-6 and TNF expression. Cell. 2010;140(2):197–208.
179. Sansone P et al. IL-6 triggers malignant features in mammospheres from human ductal breast carcinoma and normal mammary gland. J Clin Invest. 2007;117(12):3988–4002.
180. Iliopoulos D, Hirsch HA, Struhl K. An epigenetic switch involving NF-κB, Lin28, Let-7 microRNA, and IL6 links inflammation to cell transformation. Cell. 2009;139(4):693–706.
181. Miller WR. Aromatase and the breast: regulation and clinical aspects. Maturitas. 2006;54(4): 335–41.
182. Baird DT, Uno A, Melby JC. Adrenal secretion of androgens and oestrogens. J Endocrinol. 1969;45(1):135–6.
183. van Landeghem AAJ et al. Endogenous concentration and subcellular distribution of androgens in normal and malignant human breast tissue. Cancer Res. 1985;45(6):2907–12.
184. Bernstein L, Ross RK. Endogenous hormones and breast cancer risk. Epidemiol Rev. 1993; 15(1):48–65.
185. Kaaks R et al. Postmenopausal serum androgens, oestrogens and breast cancer risk: the European prospective investigation into cancer and nutrition. Endocr Relat Cancer. 2005; 12(4):1071–82.
186. Bardin A et al. Loss of ERβ expression as a common step in estrogen-dependent tumor progression. Endocr Relat Cancer. 2004;11(3):537–51.
187. Kaaks R, Lukanova A, Kurzer MS. Obesity, endogenous hormones, and endometrial cancer risk: a synthetic review. Epidemiol Biomarkers Prev. 2002;11(12):1531–43.
188. Trayhurn P, Wang B, Wood IS. Hypoxia in adipose tissue: a basis for the dysregulation of tissue function in obesity? Br J Nutr. 2008;100(2):227–35.
189. Chaffer CL, Weinberg RA. A perspective on cancer cell metastasis. Science. 2011;331(6024): 1559–64.

Chapter 6
Animal Models to Study the Interplay Between Cancer and Obesity

Amitabha Ray and Margot P. Cleary

Abstract Overweight and/or obesity are known risk factors for many cancers and are associated with poor prognosis. Evidence for this relationship has primarily been obtained from epidemiological studies with in vitro studies characterizing potential pathways that help explain the pathological role of obesity in malignancies. Animal models provide the opportunity to more completely understand disease mechanisms and intervention strategies associated with obesity and tumorigenesis. The most widely used obese animal models result from either genetic defects or consumption of high-fat diets. Genetically obese animals used in cancer research include yellow obese mice, leptin and leptin receptor-deficient mice, and the Zucker rat. Goldthioglucose-induced obesity has occasionally been used as has been ovariectomized animals. A number of studies using rodents have explored the relationship and mechanisms of obesity and the development of mammary tumors. Additional studies have evaluated the effect of obesity in colon, skin, and prostate cancer models. These studies have provided insights into the role of body weight and tumorigenesis. However, more appropriate obesity models will be important in continuing to understand the factors associated with body weight's impact on the development of cancer and to assist in providing pharmaceutical and nutritional interventions for cancer prevention and treatment.

6.1 Introduction

Obesity is recognized as a risk factor for a number of human malignancies. In addition, body weight status at the time of cancer diagnosis influences progression of the disease, responses to therapy, disease-free survival, and/or death. These data have

A. Ray • M.P. Cleary (✉)
The Hormel Institute, University of Minnesota, 801 16th Avenue NE, Austin, MN 55912, USA
e-mail: mpcleary@hi.umn.edu

M.G. Kolonin (ed.), *Adipose Tissue and Cancer*, DOI 10.1007/978-1-4614-7660-3_6,

primarily been obtained from epidemiological studies of either cross-sectional or prospective design. A number of reviews have addressed these various aspects of the relationship of body weight status and cancer in humans [1–3].

To explain the role obesity as a cancer risk factor attempts have been made to identify serum factors associated with elevated body weight that may play roles in the development and/or progression of these malignancies. In particular, for hormone-dependent cancers, there have been determinations that growth factors such as estrogen, leptin, insulin, and insulin-like growth factor-I (IGF-I) may mediate the effect of obesity on cancer. This is supported by numerous in vitro studies highlighting the impact of these individual growth factors on cell proliferation, angiogenesis, and apoptosis on human cancer cell lines. Recently, the role of inflammation resulting from obesity has also been under consideration for how this factor impacts tumorigenesis [4, 5].

What has been lacking in evaluating the complexities of how obesity impacts cancer has been identification and acceptance of animal models, which would help to connect the human observations with the in vitro studies. Animal models provide the opportunity to undertake experiments, which can be carried out in a reasonable time frame and could evaluate effects of body weight at specific time points and also evaluate various interventions. Here, we highlight publications that address this issue and make suggestions as to what types of studies may be considered in the future to address these important areas as the obesity epidemic affects populations worldwide.

6.2 Animal Models of Obesity

First, a brief overview of the obesity models that are most frequently used is presented. The most widely used animal models that have been used in cancer-related studies have been those resulting from either genetic defects or from consumption of diets with increased fat content.

With respect to mouse models of genetic obesity, the most widely used in cancer studies have been $Lep^{ob}Lep^{ob}$ (originally termed *obob*), $Lepr^{db}Lepr^{db}$ (originally termed *dbdb*), and yellow obese mice strains all of which developed from spontaneous mutations. The genetic defects resulting in obesity of *Lep* and *Lepr* mice were discovered in the mid-1990s. $Lep^{ob}Lep^{ob}$ mice do not produce the adipokine, leptin [6], while the $Lepr^{db}Lepr^{db}$ mice have a defect in the leptin receptor (OB-R) and manifest high circulating levels of leptin [7]. For both strains, homozygous recessive mice become noticeably obese at a young age and also develop hyperinsulinemia and insulin resistance. As mentioned above, another strain of genetically obese mice, which has been used in cancer studies, is the (A^{vy}) yellow obese mouse. In this strain, mutations in the *agouti* gene cause ubiquitous expression of agouti protein that has appetite stimulating effects, and the defect is associated with hyperinsulinemia [8]. Several genetically obese rat strains, Zucker and Corpulent, which both have leptin receptor defects have been used in a few cancer studies.

Diet-induced obesity is frequently employed in both rats and mice to study elevated body weight. In most rodent studies, the fat content of the diets is increased from the usual 10 % of total calories to 30 % or greater up to 60 %. The response of the animals to weight gain can be dependent upon the amount of dietary fat as well as strain and sex. This will be addressed in specific studies.

Another technique for inducing obesity include surgical removal of the ovaries to induce weight gain and/or postmenopausal like state. Also, goldthioglucose (GTG) treatment, which damages the hypothalamus resulting in overeating, has been used in several studies.

6.3 Breast/Mammary Cancer

In humans, elevated body weight and/or weight gain have been implicated as risk factors for the development of postmenopausal breast cancer [9–11]. Interestingly, even before this was a widely accepted relationship in humans there had been interest in body weight status with the development of mammary tumors as evidenced by studies in rodents dating back many decades. Here, we present results from a number of studies focused on this relationship.

The development of either spontaneous or chemically induced mammary tumors in A^{vy} mice has been reported. For example, spontaneous mammary tumor incidence [murine mammary tumor virus (MMTV)] reached essentially 100 % for both breeding lean and virgin yellow obese female mice at 8 months of age, while the virgin lean mice did not reach 100 % incidence until 16 months of age [12, 13]. Also, hyperplastic alveolar nodules (HAN) were detected at younger ages in virgin yellow obese mice compared to their lean counterparts, i.e., at 6–7 months of age 64 % of yellow obese mice exhibited HAN compared to 31 % for lean mice [14]. Further there were no detectable mammary tumors in lean mice compared to a 20 % incidence in the yellow obese mice.

Another approach to evaluate the effect of obesity resulting from the A^{vy} mutation was done using these mice on the balb/c background [15]. Mice were treated with one of two doses of 7,12-dimethylbenz[a]anthracene (DMBA) (1.5 or 6.0 mg) starting at 8 weeks of age and continuing once a week for 6 weeks. The mice were then followed for over 1 year. At the termination of the study, mammary tumor incidence was higher in the obese compared to the lean mice, 43 % vs. 33 % at the 1.5 mg dose and 86 % vs. 71 % at the higher dose of 6.0 mg. Latency was affected by body weight more than incidence in this study.

Interestingly $Lep^{ob}Lep^{ob}$ female obese mice had reduced spontaneous mammary tumor development compared to lean mice although the age of mammary tumor detection was shorter, 10.7 vs. 17.6 months of age [16]. When this mouse strain was crossed with transgenic mice that overexpress human transforming growth factor-alpha (TGF-α), no mammary tumors were detected following the mice for up to 2 years [17]. In contrast, lean wild-type mice or those with one copy of the obese gene had mammary tumor incidence rates of 50 % and 67 %, respectively. A second study was done whereby this same transgenic mouse strain, MMTV-TGF-α, was crossed with leptin receptor-deficient *Lepr^db^Lepr^db^* mice [18]. As described for the leptin-deficient *Lep^ob^Lep^ob^* mice, the leptin receptor-deficient mice also did not develop mammary tumors, while the wild-type and heterozygous mice had incidence rates of 69 % and 82 %, respectively.

Several studies used obese rat strains with leptin receptor defects. Zucker rats did not develop *N*-methyl-*N*-nitrosourea (MNU)-induced mammary tumors [19] but did develop mammary tumors at a greater incidence and shortened latency when administered DMBA and followed for ~150 days [20]. Corpulent rats had 100 % incidence of mammary tumors 112 days after administration of DMBA, while the lean mice only reached an incidence level of 20 % at this age [21]. However, in our laboratory, we were unable to induce mammary tumors with DMBA in obese Zucker rats using the same protocol and diet described for the Corpulent rats (unpublished data). These studies in conjunction with a series of in vitro studies have led to the conclusion that leptin may be an important growth factor for mammary tumor development. The fact that in some cases mammary tumors developed despite defects in the leptin receptor may be a consequence of permissiveness of the leptin receptor in different colonies of rats as well as the very high serum leptin levels characteristic of these models, which may be able to override receptor defects.

Studies evaluating mammary tumor development using mice made obese by injection of goldthioglucose (GTG) have been reported. For example, C3H mice were injected at 2–3 months of age with GTG (10 mg), which resulted in 80 % of the survivors developing obesity [22]. The 50 % incidence rate for development of spontaneous mammary tumors for the GTG obese mice was reached at 295 days at which time the incidence rate was only 19 % for the lean control mice. The control lean group did not reach the 50 % incidence level until they were 354 days of age. In another study, ovariectomized GTG obese mice were implanted with T47-D human breast cancer cells [23], an estrogen receptor positive cell line. In in vitro assays, this cell line was reported to increase proliferation with exposure to leptin [24, 25]. In the first study, obese mice reached on average 50 g compared to 33 g for lean mice. Mice were not only implanted with the cells but also with estradiol pellets. Interestingly, mice did not exhibit any tumor growth from the breast cancer cell inoculations. In a second experiment, some mice were implanted with placebo pellets rather than estradiol. In this environment, the obese mice without estradiol had 100 % incidence of mammary tumors as did 50 % of the nonobese mice, while obese with estradiol again had 0 % incidence. In both experiments, serum leptin levels of the GTG obese mice were elevated sixfold or more above levels of lean mice. In contrast, in an earlier study using C3H mice that were castrated at 12 weeks of age and 2 weeks later given GTG (7.5 mg) spontaneous mammary tumor development was reduced to 16 % compared to 44 % for obese intact mice, while it was 34 % for lean control mice but no tumors developed in the control castrated mice [26]. These findings demonstrate the complex nature of interactions of obesity with other factors including sex hormones.

High-fat diets have been used in both rats and mice to assess effects on mammary tumor development as reviewed previously [27, 28]. Here, we focus on studies where body weight increased and obesity resulted from the consumption of the high-fat diets. In several studies, mice were fed a moderately high-fat diet and then divided into groups based upon whether or not they increased body weight [29–31]. Thus, all the mice were fed the same diet. In two cases, MMTV-TGF-α mice on C57BL6 background were used and it was found that the mice defined as Obesity-Prone had mammary tumors detected at an earlier age compared to the nonobese counterparts [30, 31]. Also the Obesity-Prone mice had some tumors, which were classified as high-grade adenocarcinomas. This same approach was used in MMTV-neu mice (FVB/N background), which develop estrogen receptor negative tumors, but in this case there was not an effect of body weight on mammary tumor development [29]. This was confirmed in a recent study where the focus was more on the high-fat diet per se [32].

Another approach to obtain mice with different body weights was to feed diets with differing fat contents [33]. The mice classified as obese were fed a high-calorie diet with 5.2 kcal/g compared to the Overweight group fed 3.8 kcal/g (low-fat diet fed ad libitum) and mice classified as lean were 30 % calorie restricted. Further to mimic premenopausal vs. postmenopausal status, some mice were ovariectomized at 5 weeks of age. These C57BL6 mice were inoculated with mammary tumor cells from syngeneic Wnt-1 transgenic mice 25 weeks after being placed on the diets. After 44 days in both ovariectomized and non-ovariectomized mice, tumor growth was highest in the obese mice with overweight in the middle and the lean mice had the smallest tumors. Serum insulin, tissue plasminogen activator inhibitor (t-PAI), and IGF-I levels were elevated in response to the body weight increases, while there was little effect on adiponectin and resistin levels.

One of the primary hypothesis as to how obesity is linked to increased breast cancer risk is due to elevated aromatase activity in adipose tissue, which converts circulating androgens to estrogen [34, 35]. Recently, this aspect of breast cancer etiology was studied in a high-fat diet ovariectomized mouse model [36]. The high-fat diet contained 60 % fat by calories in comparison to 10 % for the control mice. Mice were ovariectomized after weaning and the diets started at 5 weeks of age. Mice were followed for 10 additional weeks. There was a graded effect on body weight with the high-fat ovariectomized mice weighing the most followed by high fat, low-fat ovariectomized and the low-fat intact mice were the lightest. It was noted that various measurements associated with inflammation were increased in the heavier mice, particularly the high-fat plus ovariectomized group. Assessment measurements included determining the number of inflammatory foci in the mammary gland as well as in visceral fat, measuring prostaglandin E2 levels in mammary glands and aromatase activity. Further, it was determined that proinflammatory mediators and aromatase activity were affected differently in adipocytes vs. stromal vascular fractions of the mammary gland. Although mammary tumor development per se was not presented now that the model has been established this is clearly a next step.

Several studies have addressed the role of body weight in mammary tumor development in dogs. In one study, leanness was associated with reduced risk of the disease, but obesity did not enhance it [37]. However, in another report, obesity at 1 year of age was associated with a significantly increased risk of mammary tumors [38]. Dogs with mammary tumors also were obese at the time of diagnosis, but the results did not reach statistical significance ($p<0.06$).

Table 6.1 Effects of body weight on mammary tumor development in mice

Obesity model	Cancer etiology	Outcome	References
Genetic-yellow obese A^{vy}	Spontaneous	Obese mice MTs at younger age than lean mice	[12–14]
Genetic-yellow obese A^{vy}	DMBA	Obese mice shortened latency and increased incidence	[15]
Genetic-$Lep^{ob}Lep^{ob}$	Spontaneous	Obese mice shortened latency but incidence reduced	[16]
Genetic-$Lep^{ob}Lep^{ob}$	Transgenic-TGF-α	No tumors	[17]
Genetic-$Lepr^{db}Lepr^{db}$	Transgenic-TGF-α	No tumors	[18]
Diet-induced (33 % fat diet-mice divided by weight gain	Transgenic-TGF-α (C57BL6)	Shortened latency More palpable tumors Some high-grade adenocarcinomas	[30, 31]
Diet-induced (33 % fat diet-mice divided by weight gain	Transgenic-neu (FVB/N)	Little effect of obesity on latency or incidence	[29, 32]
GTG-induces	Spontaneous C3H	Latency until 50 % incidence shortened in obese mice	[22]
GTG-induced	Spontaneous C3H	Obese intact mice increased incidence compared to lean mice, but ovariectomy protected obese and lean mice	[26]

A summary of the findings related to animal models and the development of mammary tumors is presented in Table 6.1. The studies highlighted are those that focus on spontaneous, transgenic, or chemically induced etiologies as these best reflect the effect of body weight on development whereas studies using inoculated cell lines better reflect prognosis. In general, the majority of the findings support a role for body weight in the development of mammary tumors. However, the fact that young animals as opposed to mature ones are usually used does detract from the possible significance of the results since the major impact of body weight in humans appears to be on the development of peri- and postmenopausal breast cancers. Thus, the task in the future will be to utilize physiologically relevant models to focus on mechanisms of action. Overall potential biological mechanisms, which may participate in the association between obesity and breast cancer pathologies, include systemic low-grade inflammation, metabolic, and endocrine factors.

6.4 Colon and Intestinal Cancers

Obesity has been associated with the development of cancers of the gastrointestinal tract including the colon [39–41]. In attempts to understand this relationship, the impact of elevated body weight on experimental gastrointestinal and/or colon cancer including carcinogen-induced and transgenic disease has been investigated in several obesity rodent models.

With respect to carcinogen induction of colon cancer administration of azoxymethane (AOM) to genetically obese *Lepr*^{db}*Lepr*^{db} mice from 4–14 weeks of age resulted in increased multiplicity (double the number) of premalignant lesions compared to the number found in both lean wild-type and heterozygous mice with one copy of the *Lepr*^{db} gene [42]. The mice were all fed the same amount of food, which resulted in the obese mice weighing the same as the lean mice at the termination of the experiment, but they still had significantly higher epididymal fat pad weights and serum insulin and leptin levels than did lean mice. Upregulation of OB-R and IGF-I receptor (IGF-IR) was also noted for the obese mice in cytoplasm of dysplastic and neoplastic colon tissues.

The response of *Lepr*^{db}*Lepr*^{db} as well as *Lep*^{ob}*Lep*^{ob} obese mice to the development of AOM-induced aberrant crypt foci (ACF) formation was also examined [43]. Mice were treated with 4 weekly injections of AOM starting at 7 weeks of age and then followed for 14 weeks until 25 weeks of age. At this time, obese mice had almost three times the number of total ACF/per colon than did control lean mice. Similar results were obtained following MNU administration. These results appeared to be independent of leptin as both the *Lepr*^{db}*Lepr*^{db} and lean wild-type control mice were found to have similar OB-Ra mRNA expression levels, while the *Lep*^{ob}*Lep*^{ob} mice had serum leptin levels, which were only 30 % of the wild-type mice. Nevertheless, the investigators had insufficient colonic mucosal protein to determine CYP2E1 activity, which is induced by leptin and the key enzyme in the metabolic activation of AOM. These findings suggest that obesity-related phenomena, such as hyperinsulinemia, insulin resistance, and hyperglycemia, may be responsible for the increased susceptibility to ACF formation in obese *Lepr*^{db}*Lepr*^{db} and *Lep*^{ob}*Lep*^{ob} mice. Moreover, leptin may promote the growth of colon cancer cells after malignant transformation from the precancerous stage.

Several studies used *Lep*^{ob}*Lep*^{ob} mice treated with AOM to assess the chemopreventive effect of navy beans. For example, *Lep*^{ob}*Lep*^{ob} mice that consumed navy beans had reduced dysplasia, adenomas, and adenocarcinomas compared those that did not [44]. A second study by this same research team evaluated the role of navy beans in ameliorating inflammation in association with the anticancer effect. No lean mice were evaluated; however, in either of these studies so the impact of obesity per se on colon cancer was not determined. *Lepr*^{db}*Lepr*^{db} obese mice treated with AOM had increased number of ACF as well as an increase in β-catenin-accumulated crypts (BCAC) compared to lean mice [45]. The potential chemopreventive effect of citrus auraptene was also assessed in this study. This compound significantly decreased the number of ACF and BCAC found in both lean and obese mice. In another study, this research group investigated the response to flavonoid compounds on preneoplastic colon lesions resulting from AOM administration only in the obese *Lepr*^{db}*Lepr*^{db} mice [46]. The mice had significant reductions in ACF and BCAC following treatment with all three compounds, nobiletin, chrysin, and quercetin. In a third study also using *Lepr*^{db}*Lepr*^{db} mice given AOM, pitavastatin treatment was used to determine its impact on preneoplastic colon lesions [47]. Serum factors affected by pitavastatin included reductions in cholesterol and leptin and an increase in adiponectin compared to mice not receiving this drug.

Obese Zucker rats with a similar genetic defect as the *Lepr*^{db} mice were used to assess the effect of AOM on ACF development [48]. Three experimental groups were compared, Obese-ad libitum-fed, Lean-ad libitum-fed, and Obese-energy-restricted (20–25 % reduction with similar intake to Lean-ad libitum-fed). The rats were followed for 8 weeks and then their colons were removed. Although the total number of ACF was the same in all groups, the Obese-ad libitum-fed rats had higher numbers of advanced ACF than did lean rats. In addition, energy restriction in obese rats reduced the ACF number. Results of an experiment presented in abstract form [49] also indicated that obese Zucker rats had macroscopic tumors present following AOM administration (7/9) compared to none in lean rats (0/10). Further, obese rats had more than twice the number of aberrant foci and aberrant crypts. In another study, obese female Zucker rats developed colon cancer following administration of MNU with a 13 % incidence compared to 0 % incidence in lean rats [19].

AOM was used in mice with the gastrin gene knocked out (GAS-KO) that interestingly had increased body weight as they aged as well as increased abdominal body fat as assessed by MRI and calculation of lipid-to-water ratio by NMR [50]. Plasma leptin and insulin levels were elevated in GAS-KO compared to wild-type mice. It is noteworthy that roughly 20 % of the GAS-KO mice did not gain excess weight and had body weights similar to those of wild-type mice. Following AOM-treatment, the GAS-KO obese mice had significantly increased number of ACF compared to the GAS-KO lean mice as well as the wild-type mice.

In an additional study the effect of AOM in another obese model was recently published [51]. Obese mice, KK-A^{y} (derived from A^{VY} mice), had eight times more colorectal ACF at 13 weeks of age than did the corresponding lean mice. Following follow-up until 19 weeks of age, KK-A^{y} mice had a 100 % tumor incidence (adenomas plus adenocarcinomas) compared to only 12 % for lean mice. In addition, tumor multiplicity was far greater in the obese mice. Obesity was characterized by higher serum interleukin-6 (IL-6) and leptin levels at both ages.

The *Apc*-mouse develops intestinal tumors in a similar fashion as humans who develop familial adenomatous polyps [52]. Genetic obesity resulting from cross-breeding *Apc*$^{1638N/+}$ mice with *Lepr*^{db} mouse strain impacted tumor development [53]. At 6 months of age, these obese mice had an increased number of intestinal tumors compared to the nonobese *Apc* mice and the obese mice were also found to have gastric and colon tumors. In a study conducted by Ding et al., obese A^{y}/*Apc*$^{Min/+}$ mice and their lean *Apc*$^{Min/+}$ counterparts were fed diets containing low, medium, or high-calcium levels and followed until they were 90 days old [54]. At the low- and medium-calcium levels, average numbers of tumors per mouse were similar for both obese and lean groups. However, at the high-calcium level, the lean mice had 70 % higher tumor development than the lower calcium groups. Deficiency of adipose tissue derived protective factor(s) in lean *Apc*$^{Min/+}$ mice fed a high calcium diet may promote tumorigenesis.

Subcutaneous growth of a colon cancer cell line (MC38) was assessed in both male [55] and female [56] mice with dietary-induced obesity resulting from intake

Table 6.2 Effect of obesity on the development of intestinal/colon cancer in mice

Obesity model	Cancer etiology	Outcome	References
*Lepr^db^Lepr^db^*	Carcinogen-AOM	Increase in premalignant lesions compared to lean mice	[42]
*Lepr^db^Lepr^db^* *Lep^ob^Lep^ob^*	Carcinogen- AOM or MNU	Increased ACF compared to lean mice	[43]
Gastrin knockout	Carcinogen-AOM	Increased ACF in KO mice that were obese compared to KO mice that did not become obese-increased leptin and insulin levels	[50]
KK-A^y	Carcinogen-AOM	More ACF at 13 weeks of age, and at 19 weeks of age, 100 % tumor incidence vs. 12 % for lean mice	[51]
*Lepr^db^Lepr^db^*	$APC^{min/-}$	More colon and intestinal tumors than lean mice at 6 months of age	[53]
*Lepr^db^Lepr^db^*	$APC^{min/-}$	No effect of obesity at two low levels of Ca in diet. At high Ca more tumors in lean mice	[54]

of a 60 % fat calorie diet. Male mice were fed the high-fat diet for 12 weeks and then inoculated with the MC38 cells and followed for an additional 5 weeks [55]. Body weights of the high-fat fed mice were 20 % higher than those fed a low-fat diet and tumor weight was doubled. Serum leptin, IGF-I, and insulin were all increased in the obese male mice. Female mice were fed the high-fat diet from 6 weeks of age and obesity was characterized by elevated serum insulin and leptin levels and in some cases with elevated IGF-I [56]. Tumor volumes were increased as well as the number of mice with tumors in the obese group. In a follow-up study, this obesity protocol was used in IGF-I-deficient mice [57]. In this case, obesity no longer impacted tumor growth of the MC38 colon cancer cells although leptin levels remained quite high. The effect of obesity on metastasis of the colon cells to the liver was also determined with the cancer cells delivered through the intrasplenic/portal route and livers examined 2 weeks later. In this model, tumor weights were doubled in the obese mice.

The impact of diet-induced obesity was investigated in rats treated with 1,2-dimethylhydrazine to induce colon tumors [58]. The rats were fed a 33 % fat diet and those that were in the upper 25 % of weight gain were considered obese and were compared to lean rats with the lowest 25 % weight gain. After 10 weeks on the diet, the rats were administered the carcinogen and then followed for an additional 12–14 weeks. Serum leptin and insulin were higher in the obese rats, while adiponectin was not affected. There was not a significant difference in ACF between the two body weight groups.

Overall, colon and intestinal cancers have been investigated with respect to the role of body weight and its impact on the development of the disease is summarized in Table 6.2. Body weight has also been reported to affect progression as evidenced by cell inoculation experiments. However, in contrast to many studies focused on mammary tumor development in a colon cancer, the findings suggest that leptin does not play a role in this type of cancer, whereas serum insulin and IGF-I appeared to have bigger impact as growth factors.

6.5 Liver Cancer

In humans, obesity is associated with increased incidence and/or mortality of liver cancers [59, 60]. In addition, there is mounting evidence that obesity and related metabolic complications are associated with different liver pathologies such as non-alcoholic fatty liver disease (NAFLD), nonalcoholic steatohepatitis (NASH), and cirrhosis, which may be precursors of hepatocellular carcinoma (HCC).

There were a number of early studies, which reported that genetic obesity in A^{vy} [12] and $Lep^{ob}Lep^{ob}$ mice was associated with increased liver cancer [16]. Also, tumors developed at a younger age in obese compared to lean mice. Further, it was reported that GTG-induced obesity increased spontaneous appearance of heptatomas [61].

More recently, the effects of obesity on liver malignancies in the $Lep^{ob}Lep^{ob}$ mouse model have been published. For example, adult $Lep^{ob}Lep^{ob}$ mice had an increased liver relative to body mass that was not simply due to elevated lipid storage [62]. This hepatomegaly was accompanied by significant increases in hepatocyte proliferative activity with concomitant inhibition of hepatocyte apoptosis. These findings led the authors to conclude that obesity-related metabolic abnormalities not liver cirrhosis is an important component of the relationship of obesity and liver cancer. In another study, obesity, type 2 diabetes, hyperlipidemia, and hyperinsulinemia were found in a mouse strain generated by backcross mating of the $Lep^{ob}Lep^{ob}$ mouse strain with the fatty liver Shionogi (FLS) mouse that develops spontaneous fatty liver [63]. The NASH-like lesions, including multifocal mononuclear cell infiltration and clusters of foamy cells, were observed at a younger age in FLS-$Lep^{ob}Lep^{ob}$ mice than in nonobese FLS mice. Moreover, FLS-$Lep^{ob}Lep^{ob}$ mice developed multiple hepatic tumors including adenomas and HCC following the appearance of steatohepatitis.

In other studies, high-fat diets were used to increase body weight and examine its effect on liver cancer. Hill-Baskin et al. [64] found that high-fat diet fed C57BL/6J male mice became obese, were hyperinsulinemic and susceptible to NASH and HCC in comparison to mice fed a low-fat diet. Further, mRNA profiles of HCC vs. tumor-free liver samples showed involvement of two signaling networks, one centered on Myc and the other on NF-κB. In contrast, there was little effect of the high-fat diet on any of these factors in A/J mice. In a study using mice with liver-specific inactivation of the NF-κB essential modulator gene ($NEMO^{L-KO}$), Wunderlich et al. [65] demonstrated that blocking hepatic NF-κB by disrupting NEMO in liver parenchymal cells prevented high-fat diet-induced insulin resistance, but high-fat feeding resulted in exacerbated inflammation, apoptosis, and steatosis in $NEMO^{L-KO}$ mice, ultimately leading to increased hepatic tumorigenesis. In addition, $NEMO^{L-KO}$ mice fed a high-fat diet exhibited a significant increase in tyrosine phosphorylation of signal transducer and activator of transcription 3 (STAT3) and serum IL-6 levels compared with controls.

Carcinogen-induced hepatic tumors have also been used to evaluate the effect of body weight on their development. In one study, both dietary and genetic obese mice were used to assess the effect of body weight on liver tumors that developed following diethylnitrosamine (DEN) administration [66]. Both dietary-induced obese (high-fat diet) and genetically obese (*Lep*^{ob}*Lep*^{ob}) mice had increased tumor number, size, and incidence compared to low-fat fed lean mice. Another aspect of this work included use of IL-6 deficient (*IL6*$^{-/-}$) and tumor necrosis factor receptor-deficient (TNFR1$^{-/-}$) mice fed either low-fat or high-fat diets and treated with DEN. The *IL6*$^{-/-}$ mice had reduced liver tumor development, which was not affected by higher body weight. Similar results were found for the TNFR1$^{-/-}$ mice. Also the majority of these mice fed the high-fat diet developed HCC without using the tumor promoter phenobarbital. In contrast, mice on the low-fat diet did not develop any HCC unless treated with phenobarbital after administration of DEN. Further, the obese mice displayed a substantial increase in STAT3 phosphorylation and elevated activities of c-Jun N-terminal kinase (JNK) and extracellular signal-regulated kinase (ERK) as well as enhanced expression of IL-6, TNF, and IL-1β. Results supported that both IL-6 and TNF signaling via its type 1 receptor (TNFR1) were important for lipid accumulation in the liver (hepatosteatosis) and fat-induced liver inflammation (steatohepatitis), which together defined NAFLD, a condition that greatly increases the risk of HCC development.

In a cohort of obese Swiss-Webster mice, a high prevalence of diabetes was noticed among males and many of them had grossly visible spontaneous hepatic tumors. In contrast, liver tumors were not detected in females or nondiabetic males [67]. Interestingly, Stauffer et al. [68] used transposon technology to deliver individual or combinations of genes including activated AKT and β-catenin (AKT/CAT) and MET/CAT to livers in FVB/N and C57/BL6 mice. The AKT/CAT and MET/CAT combinations induced microscopic tumor foci within 4 weeks, whereas no tumors resulted from delivery of AKT, MET, or CAT alone. Primary MET/CAT tumors emerged directly as frank HCC, whereas primary AKT/CAT tumor cells were steatotic (fatty) hepatocellular adenomas, which progressed to HCC upon in vivo passage. The investigators concluded that lipid metabolic pathways in hepatocytes and subsequent obesity-linked HCC development were associated with coactivation of the AKT and CAT pathways.

Chemical carcinogen-induced liver cancer has also been investigated in relation to body weight status. For example, diet-induced obesity (DIO) was generated by feeding C57BL6 male mice a high-fat diet beginning at 4 weeks of age; and a group of 28-week-old DIO mice were treated with the bioactivation-dependent DNA-reactive hepatocarcinogen 2-acetylaminofluorene (AAF). A similar protocol was applied to control lean mice maintained on a low-fat diet. Hepatocellular proliferation, but not hepatocarcinogen bioactivation, was identified in livers of DIO mice, which could contribute to their susceptibility to hepatocarcinogenesis [69]. In a study conducted by Shimizu et al. [70], obese hyperinsulinemic male *Lepr*^{db}*Lepr*^{db} mice were administered DEN to induce liver cancer and the chemopreventive effect of 0.1 % epigallocatechin gallate (EGCG) was determined. Even though mice were followed for 34 weeks, the effect of the intervention was minimal. In this study, 1/10 mice in each group had HCC although mice treated with DEN alone had much higher incidence of adenoma (70 % vs. 10 %). Serum insulin, IGF-I, and IGF-II

Table 6.3 Effect of obesity on the development of liver cancer in mice

Obese model	Cancer etiology	Outcome	References
A^{vy}	Spontaneous	Increased incidence in obese vs. lean Shortened latency	[12]
Lep^{ob}Lep^{ob}	Spontaneous	Increased incidence in obese vs. lean Shortened latency	[16]
GTG-induced	Spontaneous	Increased incidence	[61]
Lep^{ob}Lep^{ob} crossed with FLS	Spontaneous	Obese mice developed NASH-like symptoms at a younger age and tumors compared to FLS normal weight mice	[63]
Diet-induced C57BL6 (58 % fat)	Spontaneous	Obese mice—NASH and HCC	[64]
Diet-induced C57BL6 (59 % fat)	Carcinogen-DEN	Obese mice—increased incidence and tumor number and size	[66]
Lep^{ob}Lep^{ob}	Carcinogen-DEN	Obese mice—increased incidence and tumor number and size	[66]
Diet-induced $IL6^{-/-}$, diet-induced $TNFR1^{-/-}$ (59 % fat)	Carcinogen-DEN	Effects of obesity blocked without IL-6 and TNFR	[66]
Obese diabetic-Swiss Webster	Spontaneous	Male mice only—increased HCC	[67]

were reduced by EGCG. The levels of free fatty acids (FFA) in serum and triglycerides in the liver were reduced and the expression levels of p-AMPK proteins were increased by the administration of EGCG.

The findings related to the effect of body weight on the development of liver cancers of different etiologies are summarized in Table 6.3. In most cases, obesity was associated with shortened latency and/or increased incidence. In a few cases, attempts were made to identify mechanisms of action.

6.6 Prostate Cancer

The role of body weight in the development of human prostate cancer has been the subject of much interest. However, the published data in general do not provide definitive information as to whether and how obesity influences either the development or progression of this disease. Some studies suggested that obesity was associated with increased risk, but others have not [71, 72]. Most recently, the focus has been on obesity and characteristics of the disease [73, 74].

There are only a few published studies that have investigated the effect of either a high-fat diet and/or obesity on prostate cancer development in mouse models. Part of the problem in investigating the impact of body weight on prostate cancer development has been the shortage of suitable animal models. The TRAMP mouse has been used to an increasing extent to assess prostate cancer development and effects of various interventions. With respect to obesity, Llaverias et al. used TRAMP-C57BL6 mice fed a Western diet with 40 % fat calories plus elevated cholesterol and compared these mice to those fed a chow diet [75]. The mice were followed until 28 weeks of age. Body and epididymal fat pad weights of the Western-diet TRAMP mice were higher than those of the chow-fed mice. Further, the Western-diet mice had more advanced disease based on histopathology of the ventral and anterior prostatic lobes. The mice fed the Western diet had 70 % moderate differentiation in ventral prostatic lobe neoplasia and 25 % invasive features in the anterior prostatic lobe neoplasia with the remaining had high-grade prostatic intraepithelial neoplasia (PIN); in contrast, the chow mice only had PIN. Lung metastasis rate was almost 70 % in the mice fed the Western diet vs. 40 % for those fed chow.

Our laboratory has also used the TRAMP mouse model to investigate the effect of a high-fat diet on the development of prostate cancer [76]. A moderately high-fat diet (33 % fat calories) was fed to the mice from 7 weeks of age and at 18 weeks of age they were divided into Obesity-Prone, Overweight, Obesity-Resistant groups and followed until 50 weeks of age. There were no significant effects of the genitourinary tract (GUT) weight and there was no effect of GUT relative to body weight among the groups. Additionally, there was no effect of body weight or diet on either age to tumor detection or age at death. Pathological analysis of GUT revealed that Low-Fat and Obesity-Resistant mice i.e., those fed the high-fat diet that remained lean that had the lowest body weights had 50 % and 46 % lesions graded as either PIN or well-differentiated adenocarcinoma, respectively. The Low-Fat mice also had 50 % moderately differentiated tumors and 0 % poorly differentiated tumors, while the Obesity-Resistant mice had 46 % moderately differentiated and 8 % poorly differentiated tumors. On the other hand, the incidence of moderately or poorly differentiated tumors was 65 % and 6 % for Overweight mice and 52 % and 19 % for Obesity-Prone mice. Thus, it appeared that the heavier mice had more severe disease than did the lighter weight mice.

A different transgenic prostate cancer model, Hi-Myc mice, was used in a recent study with the goal of studying energy modulation on disease development. At 6–8 weeks of age, mice were either placed on a 60 % fat calorie diet (DIO), fed AIN-76A (overweight control) or were calorie restricted [77]. At 6 months of age, the DIO mice weighed slightly more than the control mice and had slightly more fat mass. The incidence of poorly differentiated invasive adenocarcinomas was higher in the DIO than the control mice. Given the very different diets fed and the lack of a normal weight control group, it is hard to determine if results presented were attributable to the high-fat diet or body weight changes or to other factors.

In another study, genetically obese *Lep^ob^Lep^ob^* and *Lepr^db^Lepr^db^* mice as well as DIO mice were inoculated with RM1, a murine androgen insensitive prostate carcinoma cell line [78]. The genetically obese mice weighed significantly more than lean controls; however, DIO mice did not. Furthermore, leptin and insulin levels were not different in the DIO vs. control mice. Tumor weights were highest in *Lep^ob^Lep^ob^* and DIO mice compared to control and *Lepr^db^Lepr^db^* groups. Overall, the

Table 6.4 Effect of obesity on the development of various cancers in mice

Type of cancer	Obese model	Cancer etiology	Outcome	References
Prostate	Diet-induced Western diet (40 % fat vs. chow)	Transgenic-TRAMP	More advanced disease Metastasis 70 % obese vs. 20 % in low fat	[75]
Prostate	Diet-induced (33 % fat)	Transgenic TRAMP	More advanced disease Metastasis not affected	[76]
Prostate	Diet-induced (60 % fat)	Hi-Myc	More advanced disease	[77]
Skin	Diet-induced pellet vs. powder	UV	Obese > overweight > lean greater multiplicity Reverse for latency	[80]
Skin	*Lep^{ob}Lep^{ob}*	UVB	Obese increased inflammation and proliferation	[81]
Lymphoma-leukemia	Diet-induced (60 % fat)	Transgenic BCR/ABL	C57BL6 and AKR/j mice had shortened latency	[82]
Endometrial	Diet-induced (58 % fat)	Pten$^{+/-}$	Increased abnormalities in obese but only one actual cancer	[83]

Lepr^{db}Lepr^{db} mice had the lowest tumor growth with very little Ki-67 staining and reduced mitotic index.

In a different approach, serum from genetically obese rats was obtained and its effects on human prostate cancer cells was assessed [79]. The obese rat serum was found to have an increased effect on cell proliferation compared to serum from lean rats. Charcoal stripping of the serum muted the effect. Charcoal stripped serum had reduced concentrations of basic fibroblast growth factor (FGF-2) and vascular endothelial growth factor (VEGF) in both lean and obese samples.

Overall, the data obtained from these limited mouse studies provide a complex picture for the role of diet, obesity, and perhaps adipokines on prostate cancer development. However, only a few studies have looked at the long-term effects of body weight on the development of prostate cancer as summarized in Table 6.4. The scarcity of animal models for prostate cancer has hindered examining this relationship. For the models that do exist the aggressiveness of the disease results in their not being representative of most human prostate cancers.

6.7 Skin Cancer and Melanoma

There is some evidence that obesity may increase the risk of melanoma in human. It has been reported that heavier males but not females had increased risk of melanoma [84]. However, other studies indicated that females were also at increased risk for melanoma at higher BMI categories [85, 86]. It has also been suggested that body weight at the time of diagnosis may not reflect risk, while obesity at a younger age may be more important [87].

Despite the limited evidence from human studies, there are a number reports using animal models suggesting that obesity is a risk factor for both melanoma and nonmelanoma skin cancers. In an interesting study, obesity was induced in SKH-1 hairless mice by feeding them a powdered AIN-76A diet from 5–6 until 30 weeks of age, which resulted in these mice weighing significantly more than those fed the diet in pelleted form [80]. A third group considered to be overweight was fed the pelleted diet until 17 weeks of age followed by the powdered diet resulting in intermediate weight gain. The obese mice had a shortened latency for the development of ultraviolet (UV) irradiation-induced skin tumors and increased multiplicity compared to the control group with the overweight group having intermediate values. In another study, dietary obesity resulting from a 6 months of feeding a high-fat diet resulted in heavier mice with significantly larger tumors following the inoculation of B16F10 melanoma cells into male C57BL6 mice [88]. The obese mice also had significant increases in serum insulin, leptin, and cholesterol and reduced adiponectin levels and their tumors had higher expression levels of fatty acid synthetase, pAKT, and Cav-1.

Several studies have evaluated UVB-induced skin cancer in genetically obese $Lep^{ob}Lep^{ob}$ mice. Skin of obese $Lep^{ob}Lep^{ob}$ mice had greater susceptibility to UVB-induced inflammatory responses in comparison to the skin of wild-type C57BL6 mice [81]. This was characterized by increased expression of COX-2, PGE_2, IL-6, TNF-α, and IL-1β in skin of the obese mice. Increased proliferation in skin of the obese mice was also found as reflected by mRNA expression of PCNA and cyclin D1. In another study, using $Lep^{ob}Lep^{ob}$ mice as well as obese melanocortin receptor 4 knockout MC4R$^{-/-}$mice, Brandon et al. reported that tumors originating from B16F10 melanoma cells were much larger in obese leptin deficient than in lean mice or in pair-fed $Lep^{ob}Lep^{ob}$ mice [89]. In contrast to the leptin deficiency of $Lep^{ob}Lep^{ob}$ mice, the MC4R$^{-/-}$ mice have high levels of leptin. Tumors obtained from both obese models expressed high levels of VEGF and VEGF receptor 1 (VEGF-R1). Also, tumor weight was positively correlated with VEGF concentration and higher VEGF-R1 and VEGF-R2 protein expression was detected in melanomas of obese mice. Interestingly, when serum obtained from $Lep^{ob}Lep^{ob}$ mice was added to B16BL6 melanoma cells, there was increased invasion in comparison to serum from lean mice [90]. Serum from the obese mice contained proinflammatory substances such as resistin, IL-6, and TNF-α; and it increased matrix metalloprotease-9 (MMP9) activity and decreased the expression of E-cadherin and the metastasis suppressor gene Kiss1.

In another approach, $Lep^{ob}Lep^{ob}$ and $Lepr^{db}Lepr^{db}$ mice were used to assess the effects of obesity on metastasis of B16BL6 melanoma cells [91]. Cells were injected via tail vein and lungs examined after 14 days. In both obese mouse strains, there was a large increase in the number of colonies in the lungs. Interestingly, administration of leptin to the leptin-deficient $Lep^{ob}Lep^{ob}$ mice decreased colonization of cells in the lung suggesting a leptin-independent process.

These findings strongly suggest that body weight status may be associated with skin cancers, particularly with respect to disease progression. Only a few studies have been published on the effects of body weight on the development of the disease (Table 6.4). However, the mechanism of action is probably not mediated through elevations in serum leptin, but other factors which remain to be identified.

6.8 Other Cancers

The impact of body weight on several other cancers has been reported. With respect to pancreatic cancer, studies are only available using inoculated cells thus represent progression more than development. For example, obese *Lepr*^{db}*Lepr*^{db} and *Lep*^{ob}*Lep*^{ob} mice inoculated with PAN02 murine pancreatic cells were followed for 5 weeks and results compared to lean mice [92]. Tumor weight was highest in the *Lep*^{ob}*Lep*^{ob} mice, intermediate in *Lepr*^{db}*Lepr*^{db}, and lowest in the lean mice. In addition, tumor weights were positively related to body weight and tumors from both obese groups exhibited higher proliferation rates than those from lean mice, but there was no effect on apoptosis. It was also noted that although no metastases or deaths occurred in the lean mice, they did in both groups of obese mice. Serum adiponectin levels were similar among the groups, while insulin, glucose, and HOMA values were higher in the obese mice. In another study from this same group, White et al. [93] reported that C57BL6 mice with diet-induced obesity had increased growth of tumors from PAN02 cells with tumor weight twice that of lean mice. Tumor weight was also positively correlated with body weight. Serum leptin, insulin, and HOMA were higher in the obese mice, while there was no effect on adiponectin or glucose.

Spontaneous endometrial multiglandular foci and atypia were identified in 58 % of female PTEN$^{+/-}$ mice fed a low-fat diet, but there was no endometrial carcinoma [83]. In contrast, 67 % of female PTEN$^{+/-}$ obese mice resulting from consumption of a high-fat diet (58 % fat calories) had such abnormalities and one had endometrial cancer. Another study using genetically obese Zucker rats found no differences in morphology or histology in endometrium compared to lean rats [94]. However, it was noted that following ovariectomy and subsequent estradiol treatment that endometrium from the obese rats had significantly greater expression levels of proliferative genes.

Obesity resulted from consumption of a 60 % fat by calorie diet was used in two models of acute lymphoblastic leukemia [82]. The models were transgenic BCR/ABL mice and AKR/J mice, which develop T-cell acute lymphoblastic leukemia. In both cases, obesity shortened the time until death.

6.9 Conclusions

Overall, it appears that animal models of cancer can be influenced by body weight status. Most studies have utilized mouse models. As summarized in Tables 6.1, 6.2, 6.3, and 6.4, the development of a number of cancers is influenced by body weight status. Interestingly, with respect to mammary tumor development, body weight clearly affected disease development as is the case for human postmenopausal breast cancer estrogen sensitive tumors. Major findings include the role of obesity in shortening tumor latency and perhaps affecting aggressiveness of the disease. Other studies in liver and intestinal/colon malignancies also suggest body weight impacts the development of these malignancies. Although there has been limited interest in the role of body weight and skin cancers in humans, there are actually a large number of animal studies in this area primarily related to disease progression. Perhaps this is due to the nature of the disease models, which are widely used to test chemopreventive agents and can in the case of UVB-induced cancers be easily followed externally. In contrast, despite the effect of body weight on pancreatic cancer, only a few studies have addressed this issue. This perhaps is due to the lack of good transgenic models.

Future studies should be held to high standards with respect to selection of the appropriate cancer model. Further, and very importantly, the obesity models must be appropriate. Although the genetic obesities may be easy to use, they generally do not reflect most cases of human obesity. Although as seen in mammary tumor and colon cancer studies, their use can provide valuable insights into potentially important growth factors. Diet-induced obesity is the most representative of the human situation. But here too there can be pitfalls. In some studies, mice with diet-induced obesity were fed purified high-fat diets and their results compared to lean mice fed a chow type nonpurified diet. Even when high-fat diet fed animals have been compared to those fed low-fat diet, the different diet compositions may have effects on outcomes. Results from many of these types of studies can not only be attributed to obesity.

As detailed here, overweight and/or obesity is associated with an elevated risk of several cancers; however, it is clear that a common disease mechanism was not identified. Although the current literature hypothesizes at least three major components such as sex hormones, insulin-related pathologies, and adipokines, these components cannot explain every aspect of clinical features/disease courses. But as models improve both for obesity and various cancers, hopefully it will become easier to identify mechanisms of action for the relationship of body weight and cancer.

Acknowledgments This work was supported by NIH-NCI grant CA157012 and the Hormel Foundation.

References

1. Basen-Engquist K, Chang M. Obesity and cancer risk: recent review and evidence. Curr Oncol Rep. 2011;13:71–6.
2. Calle EE, Kaaks R. Overweight, obesity and cancer: epidemiological evidence and proposed mechanisms. Nat Rev Cancer. 2004;4:579–91.
3. Pischon T, Nöthlilngs U, Boeing H. Obesity and cancer. Proc Nutr Soc. 2008;67:128–45.

4. Harvey AE, Lashinger LM, Hursting SD. The growing challenge of obesity and cancer: an inflammatory issue. Ann N Y Acad Sci. 2011;1229:45–52.
5. Lysaght J, van der Stok EP, Allott EH, et al. Pro-inflammatory and tumour proliferative properties of excess visceral adipose tissue. Cancer Lett. 2011;312:62–72.
6. Zhang Y, Proenca R, Maffei M, et al. Positional cloning of the mouse obese gene and its human homologue. Nature. 1994;372:425–32.
7. Frederich RC, Hamann A, Anderson S, et al. Leptin levels reflect body lipid content in mice: evidence for diet-induced resistance to leptin action. Nat Med. 1995;1:1311–4.
8. Wolff GL, Roberts DW, Mountjoy KG. Physiological consequences of ectopic agouti gene expression: the yellow obese mouse syndrome. Physiol Genomics. 1999;1:151–63.
9. Cleary MP, Maihle NJ. The role of body mass index in the relative risk of developing premenopausal versus postmenopausal breast cancer. Proc Soc Exp Biol Med. 1997;216:28–43.
10. Grossmann ME, Ray A, Nkhata KJ, et al. Obesity and breast cancer: status of leptin and adiponectin in pathological processes. Cancer Metastasis Rev. 2010;29:641–53.
11. Rose DP, Vona-Davis L. Interaction between menopausal status and obesity in affecting breast cancer risk. Maturitas. 2010;66:33–8.
12. Heston WE, Vlahakis G. Influence of the A^y gene on mammary-gland tumors, hepatomas, and normal growth in mice. J Natl Cancer Inst. 1961;26:969–83.
13. Heston WE, Vlahakis G. C3H-A^{vy}- a high hepatoma and mammary tumor strain of mice. J Natl Cancer Inst. 1968;40:1161–6.
14. Wolff GL, Medina D, Umholtz RL. Manifestation of hyperplastic alveolar modules and mammary tumors in "viable yellow" and non-yellow mice. J Natl Cancer Inst. 1979;63:781–5.
15. Wolff GL, Kodell RL, Cameron AM, et al. Accelerated appearance of chemically induced mammary carcinomas in obese yellow (A^{vy}/A) (BALB/c X^{vy}) F1 hybrid mice. J Toxicol Environ Health. 1982;10:131–42.
16. Heston WE, Vlahakis G. Genetic obesity and neoplasia. J Natl Cancer Inst. 1962;29:197–209.
17. Cleary MP, Phillips FC, Getzin SC, et al. Genetically obese MMTV-TGF-α/*Lep*^{ob}*Lep*^{ob} mice do not develop of mammary tumors. Breast Cancer Res Treat. 2003;77:205–15.
18. Cleary MP, Juneja SC, Phillips FC, et al. Leptin receptor deficient MMTV-TGF-α/*Lepr*^{db}*Lepr*^{db} female mice do not develop oncogene-induced mammary tumors. Exp Biol Med. 2004;229: 182–93.
19. Lee WM, Lu S, Medline A, et al. Susceptibility of lean and obese Zucker rats to tumorigenesis induced by N-methyl-N-nitrosurea. Cancer Lett. 2001;162:155–60.
20. Hakkak R, Holley AW, MacLeod SL, et al. Obesity promotes 7,12-dimethylbenz(a)anthracene-induced mammary tumor development in female Zucker rats. Breast Cancer Res. 2005;7: R627–33.
21. Klurfeld DM, Lloyd LM, Welch CB, et al. Reduction of enhanced mammary carcinogenesis in LA/N-cp (corpulent) rats by energy restriction. Proc Soc Exp Biol Med. 1991;196:381–4.
22. Waxler SH, Tabar P, Melcher LP. Obesity and the time of appearance of spontaneous mammary carcinoma in C3H mice. Cancer Res. 1953;13:276–8.
23. Nkhata KJ, Ray A, Dogan S, et al. Mammary tumor development from T47-D human breast cancer cells in obese ovariectomized mice with and without estradiol supplements. Breast Cancer Res Treat. 2009;114:71–83.
24. Hu X, Juneja SC, Maihle NJ, et al. Leptin- a growth factor for normal and malignant breast cells and normal mammary gland development. J Natl Cancer Inst. 2002;94:1704–11.
25. Laud K, Gourdou I, Pessemesse L, et al. Identification of leptin receptors in human breast cancer: functional activity in the T47-D breast cancer cell line. Mol Cell Endocrinol. 2002; 188:219–26.
26. Waxler SH, Leef MF. Augmentation of mammary tumors in castrated obese C3H mice. Cancer Res. 1966;26:860–2.
27. Pariza MW. Fat, calories, and mammary carcinogenesis: net energy effects. Am J Clin Nutr. 1987;45:261–3.
28. Welsch CW, House JL, Herr BL, et al. Enhancement of mammary carcinogenesis by high levels of dietary fat: a phenomenon dependent on ad libitum feeding. J Natl Cancer Inst. 1990;82:1615–20.

29. Cleary MP, Grande JP, Juneja SC, et al. Effect of dietary-induced obesity and mammary tumor development in MMTV-neu female mice. Nutr Cancer. 2004;50:174–80.
30. Cleary MP, Grande JP, Maihle NJ. Effect of a high fat diet on body weight and mammary tumor latency in MMTV-TGF-α mice. Int J Obes Relat Metab Disord. 2004;28:956–62.
31. Dogan S, Hu X, Zhang Y, et al. Effects of high fat diet and/or body weight on mammary tumor leptin and apoptosis signaling pathways in MMTV-TGF-α mice. Breast Cancer Res. 2007;9:R91.
32. Khalid S, Hwang D, Babichev Y, et al. Evidence for tumor promoting effect of high-fat diet independent of insulin resistance in HER2/Neu mammary carcinogenesis. Breast Cancer Res Treat. 2010;122:647–59.
33. Núnez NP, Perkins SN, Smith NCP, et al. Obesity accelerates mouse mammary tumor growth in the absence of ovarian hormones. Nutr Cancer. 2008;60:534–41.
34. Brodie A, Lu Q, Nakamura J. Aromatase in the normal breast and breast cancer. J Steroid Biochem Mol Biol. 1997;61:281–6.
35. Siiteri PK. Adipose tissue as a source of hormones. Am J Clin Nutr. 1987;45:277–82.
36. Subbaramaiah K, Howe LR, Bhardwaj P, et al. Obesity is associated with inflammation and elevated aromatase expression in the mouse mammary gland. Cancer Prev Res (Phila). 2011;4: 329–46.
37. Sonnenschein EG, Glickman LT, Goldschmidt LT, et al. Body conformation, diet, and risk of breast cancer in pet dogs: a case-control study. Am J Epidemiol. 1991;133:694–703.
38. Alenza DP, Rutterman GR, Peña L, et al. Relation between habitual diet and canine mammary tumors in a case-control study. J Vet Intern Med. 1998;12:132–9.
39. Calle EE, Thun MJ. Obesity and cancer. Oncogene. 2004;23:6365–78.
40. Donohoe CL, Pidgeon GP, Lysaght J, et al. Obesity and gastrointestinal cancer. Br J Surg. 2010;97:628–42.
41. Freeman HJ. Risk of gastrointestinal malignancies and mechanisms of cancer development with obesity and its treatment. Best Pract Res Clin Gastroenterol. 2004;18:1167–75.
42. Hirose Y, Hata K, Kuno T, et al. Enhancement of development of azoxymethane-induced colonic premalignant lesions in C57BL/KsJ-*/db/db* mice. Carcinogenesis. 2004;25:821–5.
43. Ealey KN, Lu S, Archer MC. Development of aberrant crypt foci in the colons of ob/ob and db/db mice: evidence that leptin is not a promoter. Mol Carcinog. 2008;47:667–77.
44. Bobe G, Barrett KG, Mentor-Marcel RA, et al. Dietary cooked navy beans and their fractions attenuate colon carcinogenesis in azoxymethane-induced ob/ob mice. Nutr Cancer. 2008; 60:373–80.
45. Hayashi K, Suzuki R, Miyamoto S, et al. Citrus auraptene suppresses azoxymethane-induced colonic preneoplastic lesions in C57BL/KsJ*/db/db* mice. Nutr Cancer. 2007;58:75–84.
46. Miyamoto S, Yasui Y, Ohigashi H, et al. Dietary flavonoids suppress azoxymethane-induced colonic preneoplastic lesions in male C57BL/KsJ-*db/db* mice. Chem Biol Interact. 2010; 183:276–83.
47. Yasuda Y, Shimizu M, Shirakami Y, et al. Pitavastatin inhibits azoxymethane-induced colonic preneoplastic lesions in C57BL/KsJ-*db/db* obese mice. Cancer Sci. 2010;101:1701–7.
48. Raju J, Bird RP. Energy restriction reduces the number of advanced aberrant crypt foci and attenuates the expression of colonic transforming growth factor β and cyclooxygenase isoforms in Zucker obese (fa/fa) rats. Cancer Res. 2003;63:6595–601.
49. Weber RV, Stein DE, Kim J, et al. Obesity potentiates experimental colon cancer. Int J Obes. 1997;20:S85.
50. Cowey SL, Quast M, Belalcazar LM, et al. Abdominal obesity, insulin resistance, and colon carcinogenesis are increased in mutant mice lacking gastrin gene expression. Cancer. 2005;103:2643–53.
51. Teraoka N, Mutoh M, Takasu S, et al. High susceptibility to azoxymethane-induced colorectal carcinogenesis in obese KK-A^y mice. Int J Cancer. 2011;129:528–35.
52. Fodde R, Edelmann W, Yang K, et al. A targeted chain-termination mutation in the mouse Apc gene results in multiple intestinal tumors. Proc Natl Acad Sci USA. 1994;91:8969–73.

53. Gravaghi C, Bo J, LaPerle KMD, et al. Obesity enhances gastrointestinal tumorigenesis in APC-mutant mice. Int J Obes. 2008;32:1716–9.
54. Ding S, McEntee MF, Whelan J, et al. Adiposity-related protection of intestinal tumorigenesis: interaction with dietary calcium. Nutr Cancer. 2007;58:153–61.
55. Algire C, Amrein L, Zakikhani M, et al. Metformin blocks the stimulative effect of a high-energy diet on colon carcinoma growth in vivo and is associated with reduced expression of fatty acid synthase. Endocr Relat Cancer. 2010;17:351–60.
56. Yakar S, Nunez NP, Pennisi P, et al. Increased tumor growth in mice with diet-induced obesity: impact of ovarian hormones. Endocrinology. 2006;147:5826–34.
57. Wu Y, Brodt P, Sun H, et al. Insulin-like growth factor-I regulates the liver microenvironment in obese mice and promotes liver metastasis. Cancer Res. 2010;70:57–67.
58. Drew JE, Farquharson AJ, Padidar S, et al. Insulin, leptin, and adiponectin receptors in colon: regulation relative to differing body adiposity independent of diet and in response to dimethylhydrazine. Am J Physiol. 2007;293:G682–91.
59. Calle EE, Rodriguez C, Walker-Thurmond K, et al. Overweight, obesity, and mortality from cancer in a prospectively studied cohort of U.S. adults. N Engl J Med. 2003;348:1625–38.
60. Qian Y, Fan J-G. Obesity, fatty liver and liver cancer. Hepatobiliary Pancreat Dis Int. 2005;4: 173–7.
61. Waxler SH. Obesity and cancer susceptibility in mice. Am J Clin Nutr. 1960;8:760–6.
62. Yang S, Lin HZ, Hwang J, et al. Hepatic hyperplasia in noncirrhotic fatty livers: is obesity-related hepatic steatosis a premalignant condition? Cancer Res. 2001;61:5016–23.
63. Soga M, Hashimoto M, Kishimoto Y, et al. Insulin resistance, steatohepatitis, and hepatocellular carcinoma in a new congeneic strain of fatty liver Shionogi (FLS) mice with the *Lep*^{ob} gene. Exp Anim. 2010;59:407–19.
64. Hill-Baskin AE, Markiewski MM, Buchner DA, et al. Diet-induced hepatocellular carcinoma in genetically predisposed mice. Hum Mol Genet. 2009;18:2975–88.
65. Wunderlich FT, Luedde T, Stinger S, et al. Hepatic NF-κB essential modulator deficiency prevents obesity-induced insulin resistance but synergizes with high-fat feeding in tumorigenesis. Proc Natl Acad Sci USA. 2008;105:1297–302.
66. Park EJ, Lee JH, Yu G-Y, et al. Dietary and genetic obesity promote liver inflammation and tumorigenesis by enhancing IL-6 and TNF expression. Cell. 2010;140:197–208.
67. Lemke LB, Rogers AB, Nambiar PR, et al. Obesity and non-insulin-dependent diabetes mellitus in Swiss-Webster mice associated with late-onset hepatocellular carcinoma. J Endocrinol. 2008;199:21–32.
68. Stauffer JK, Scarzello AJ, Anderson JB, et al. Coactivation of AKT and β-catenin in mice rapidly induces formation of lipogenic liver tumors. Cancer Res. 2011;71:2718–27.
69. Iatropoulos MJ, Duan JD, Jeffrey AM, et al. Hepatocellular proliferation and hepatocarcinogen bioactivation in mice with diet-induced fatty liver and obesity. Exp Toxicol Pathol. 2013;65:451–6.
70. Shimizu M, Sakai H, Shirakami Y, et al. Preventive effects of (−)-epigallocatechin gallate on diethylnitrosamine-induced liver tumorigenesis in obese and diabetic C57BL/KsJ-*db/db* mice. Cancer Prev Res (Phila). 2011;4:396–403.
71. Amling CL. Relationship between obesity and prostate cancer. Curr Opin Urol. 2005;15: 167–71.
72. Cao Y, Ma J. Body mass index, prostate cancer-specific mortality, and biochemical recurrence: a systematic review and meta-analysis. Cancer Prev Res (Phila). 2011;4:486–501.
73. De Nunzio C, Freedland SJ, Miano L, et al. The uncertain relationship between obesity and prostate cancer: an Italian biopsy cohort analysis. Eur J Surg Oncol. 2011;37:1025–9.
74. Fowke JH, Motley SS, Concepcion RS, et al. Obesity, body composition, and prostate cancer. BMC Cancer. 2012;12:23.
75. Llaverias G, Danillo C, Wang Y, et al. A Western-type diet accelerates tumor progression in an autochtonous mouse model of prostate cancer. Am J Pathol. 2010;177:3180–91.
76. Bonorden MJL, Grossmann ME, Ewing SA, et al. Growth and progression of TRAMP prostate tumors in relation to diet and obesity. Prostate Cancer. 2012;2012:543970.

77. Blando J, Moore T, Hursting S, et al. Dietary energy balance modulates prostate cancer progression in Hi-Myc mice. Cancer Prev Res (Phila). 2011;4:2002–14.
78. Ribeiro AM, Andrade S, Pihno F, et al. Prostate cancer cell proliferation and angiogenesis in different obese mice models. Int J Exp Pathol. 2010;91:374–86.
79. Lamarre NS, Ruggieri MR, Braverman AS, et al. Effect of obese and lean Zucker rat sera on human and rat prostate cancer cells: implications in obesity-related prostate tumor biology. Urology. 2007;69:191–5.
80. Dinkova-Kostova AT, Fahey JW, Jenkins SN, et al. Rapid body weight gain incrases the risk of UV radiation-induced skin carcinogenesis in SKH-1 hairless mice. Nutr Res. 2008;28: 539–43.
81. Sharma SD, Katiyar SK. Leptin deficiency-induced obesity exacerbates ultraviolet B radiation induced cyclooxygenase-2 expression and cell survival signals in ultraviolet B-irradiated mouse skin. Toxicol Appl Pharmacol. 2010;244:328–35.
82. Yun JP, Behan JW, Heisterkamp N, et al. Diet-induced obesity accelerates acute lymphoblastic leukemia progression in two murine models. Cancer Prev Res (Phila). 2010;3:1259–64.
83. Yu W, Cline M, Maxwell LG, et al. Dietary vitamin D exposure prevents obesity-induced increase in endometrial cancer in *Pten*$^{+/-}$ mice. Cancer Prev Res (Phila). 2010;3:1246–58.
84. Shors AR, Solomon C, McTiernan A, et al. Melanoma risk in relation to height, weight, and exercise (United States). Cancer Causes Control. 2001;12:599–606.
85. Gallus S, Naldi L, Martin L, et al. Anthropometric measures and risk of cutaneous malignant melanoma: a case-control study from Italy. Melanoma Res. 2006;16:83–7.
86. Naldi L, Altieri A, Imberti G, et al. Cutaneous malignant melanoma in women. Phenotypic characteristics, sun exposure, and hormonal factors: a case-control study from Italy. Ann Epidemiol. 2005;15:545–50.
87. Dennis LK, Lowe JB, Lynch CF, et al. Cutaneous melanoma and obesity in the agricultural health study. Ann Epidemiol. 2008;18:214–21.
88. Pandley V, Vijaykumar MV, Ajay AK, et al. Diet-induced obesity increases melanoma progression: involvement of Cav-1 and FASN. Int J Cancer. 2012;130:497–508.
89. Brandon EL, Gu J-W, Cantwell L, et al. Obesity promotes melanoma tumor growth: role of leptin. Cancer Biol Ther. 2009;8:1871–9.
90. Kushiro K, Núnez NP. Ob/ob serum promotes mesenchymal cell phenotype in B16BL6 melanoma cells. Clin Exp Metastasis. 2011;28:877–86.
91. Mori A, Sakurai H, Choo M-K, et al. Severe pulmonary metastasis in obese and diabetic mice. Int J Cancer. 2006;119:2760–7.
92. Zyromski NJ, Mathur A, Pitt HA, et al. Obesity potentiates the growth and dissemination of pancreatic cancer. Surgery. 2009;146:258–63.
93. White PB, True EM, Ziegler KM, et al. Insulin, leptin, and tumoral adipocytes promote murine pancreatic cancer growth. J Gastrointest Surg. 2010;14:1888–94.
94. Zhang Q, Shen Q, Celestino J, et al. Enhanced estrogen-induced proliferation in obese rat endometrium. Am J Obstet Gynecol. 2009;200:186.e1–8.

Chapter 7
Unraveling the Local Influence of Tumor-Surrounding Adipose Tissue on Tumor Progression: Cellular and Molecular Actors Involved

Catherine Muller, Laurence Nieto, and Philippe Valet

Abstract Cells that compose the tumor stroma are associated, if not obligate, partners in tumor progression. Among the different cell types frequently found at close proximity of evolving tumors, little attention has been given to cells that compose the adipose tissue (AT) although a growing interest can be noted in recent years. AT is mainly composed of mature adipocytes that are able to secrete a large panel of bioactive molecules (adipokines) and adipose progenitors. Emerging studies clearly indicate that a bidirectional crosstalk is established between all cellular components of AT and cancer cells and that the tumor-surrounding AT contributes to inflammation, extracellular matrix remodeling as well as energy supply within the tumors. In this chapter, we present evidences showing how AT locally affects tumor progression in given types of tumors and how these results might be attractive to explain the link between obesity and the poor prognosis of some cancers. This will be preceded by the overall description of AT composition and function with special emphasis on the specificity of adipose depots, key aspects that need to be taken in account when paracrine effects of AT on tumor progression is considered.

C. Muller (✉) • L. Nieto
Université de Toulouse, UPS, 31077 Toulouse, France

Institut de Pharmacologie et de Biologie Structurale, CNRS UMR 5089,
205 route de Narbonne, BP 64182, 31077 Toulouse Cedex, France
e-mail: muller@ipbs.fr

P. Valet
Université de Toulouse, UPS, 31077 Toulouse, France

Institut National de la Santé et de la Recherche Médicale, INSERM U1048,
31432 Toulouse, France

M.G. Kolonin (ed.), *Adipose Tissue and Cancer*, DOI 10.1007/978-1-4614-7660-3_7,

7.1 Introduction

The past two decades have provided substantial evidence for the major role of the tissue local environment for tumor progression. Cancer is now considered as a tissue-based disease in which malignant cells interact dynamically with the surrounding supportive tissue, the tumor stroma, composed by multiple normal cell types such as fibroblasts, infiltrating immune cells, and endothelial cells within the context of extracellular matrix [1]. This stroma/tumor cell interaction involves constant bidirectional crosstalk between normal and malignant cells. Cancer cells usually generate a supportive microenvironment by activating the wound-healing response of the host [2]. Conversely, the stromal cells, such as for example, cancer-associated fibroblasts (CAFs) or tumor-associated macrophages (TAMs), promote tumor progression through different mechanisms including enhancement of tumor survival, growth, and spread, by secreting growth factors, chemokines, extracellular matrix (ECM) components, and ECM-modifying enzymes [3, 4].

Constituents of the tumor microenvironment can arise from two major sources: recruitment from nearby local tissue or systemic recruitment from distant tissues via circulation. Among the different cell types frequently found at close proximity of evolving tumors, little attention has been given to cells that compose the adipose tissue (AT) although a growing interest can be noted in recent years. Throughout the body, AT is mainly described as subcutaneous (i.e., superficial and deep hypodermic location) and visceral depots. Visceral adipose tissue (VAT) surrounds the inner organs and can be divided into omental, mesenteric, retroperitoneal (surrounding the kidney), gonadal, perivascular, and pericardial depots [5]. Of note, AT is also present in the breast (mammary adipose tissue or MAT) and in the bone marrow (BM). All these specific regional depots exhibit differences in structure, function, composition, and secretion profiles [6]. This is more evidently demonstrated by the links revealed between increased deep abdominal or visceral, but not peripheral, fat extent, and the metabolic disturbances associated with obesity [6]. The cellular heterogeneity of AT adds an additional degree of complexity when AT/cancer cells crosstalk is considered. AT is mainly composed of mature adipocytes that are lipid-filled cells. In addition, other cell types are present in its stroma vascular fraction (SVF); it is including adipose progenitors [(Adipose-Derived Stem Cells (ADSCs) and preadipocytes)], lymphocytes, macrophages, fibroblasts, and vascular endothelial cells [7]. All the cells from adipose tissue (including mature adipocytes) produces a large number of secretory bioactive substances, such as hormones, growth factors, chemokines, proangiogenic or proinflammatory molecules [8], which could directly affect adjacent tumors. AT is therefore an excellent candidate to influence tumor behavior through heterotypic paracrine signaling processes and might prove to be critical for tumor survival, growth, local, and distant invasion.

Emerging studies clearly indicate that a bidirectional crosstalk is established between cellular components of AT and cancer cells and that the tumor-surrounding AT contributes to inflammation [9, 10], extracellular matrix remodeling [11, 12] as well as energy supply within the tumors [13]. It is worth noting that most experimental data obtained using tumor cell lines and animal models, which were then validated in human tumors concern for most part mature adipocytes, and not fat cell precursors. These studies are of major clinical importance since, as stated in previous chapters, several epidemiological studies demonstrate positive associations between the prevalence of obesity, as judged by increased body mass index (BMI), cancer incidence, and cancer-related mortality (for review see [14]). Therefore, it is tempting to speculate that the dysfunctions of AT encountered in the obese might positively affect the deleterious crosstalk established with cancer cells. Concerning adipose precursors, studying their role in tumor progression is also highly relevant to the clinical situations. First, due to the fact that these cells have a great potential for use in regenerative medicine [15], safety of their use regarding cancer initiation and progression needs to be questioned. Second, as obesity is associated to both hypertrophy and hyperplasia of AT, it is reasonable to speculate that the increase number in adipose progenitors might favor their recruitment and act into actively growing tumors as recently proposed [16].

In this chapter, we present evidences showing how AT locally affects tumor progression in given types of tumors and how these results might be attractive to explain the link between obesity and the poor prognosis of some cancers. This will be preceded by the overall description of AT composition and function with special emphasis on the specificity of adipose depots, key aspects that need to be taken in account when paracrine effects of AT on tumor progression is considered.

7.2 Adipose Tissues: Different Depots, Different Functions, and Different Impact on Human Health

AT is among the very few vertebrate tissues to have escaped until recently the attention of comparative anatomists. Fat depots from different region of the body have different incidence in pathology because they display distinct functional and structural properties in terms of energy metabolism and bioactive molecule (adipokines) release as well. Regional heterogeneity plays a central role in mammalian AT homeostasis. Such a physiological specialization is adaptive and functionally related with biochemical properties. The lack of universally accepted definition of nomenclature of fat depots led to a global view of three major localizations of lipid accumulation: the brown, the white visceral (VAT), and the white subcutaneous (SAT) adipose tissues. Although recent studies described its presence in humans [17], the brown adipose tissue (BAT) is mainly involved in heat production for hibernating animals and small rodents; it will not be further described in the present chapter for review see [18]. The white adipose tissues (WAT) are artificially separated in visceral fat localized in the abdominal cavity-mediastinum and subcutaneous fat in the hypodermis. While subcutaneous fat is regarded as responsible of thermal insulation and is less metabolically active, excess of visceral fat is associated with a rise in insulin resistance, type 2 diabetes, and cardiovascular diseases, risks partly related to an impaired nonesterified fatty acid (NEFA) metabolism [19]. In addition, accumulation of macrophages and immune cells in VAT of obese patients lead to the production of high levels of proinflammatory factors. These inflammatory-related events also contribute to obesity-related diseases [20].

SAT and VAT exhibit both metabolic and secretory differences that explain their differential effect on human health. Metabolic differences between VAT and SAT are due to impairment in both storage and mobilization capacities. Insulin is the major hormone involved in storage (by stimulating lipogenesis, fatty acid (FA), and glucose uptake, FA reesterification) as well as in lipid mobilization, since it inhibits both basal and catecholamine-stimulated lipolysis. Insulin actions (antilipolysis, triglyceride synthesis, and FA reesterification) are clearly blunted in VAT when compared to SAT leading to a higher lipolysis rate, which is causative of a rise in NEFA flux to the liver and an impaired liver metabolism in the obese. At the cellular level, these differences could be explained by alterations in the insulin-signaling cascade [21]. Another major hormonal system controlling metabolic activity of AT is the catecholamines, since they are able to control the lipolytic rate of the adipocytes by either stimulating the hydrolysis of triglycerides via the activation of β-adrenergic receptors or blocking the same pathway by activating the α2-adrenergic receptors. The lipolytic response of fat cells to catecholamines is weaker in the SAT than in the VAT since the latter exhibits highest amounts of α2-adrenoceptors and the lowest β-adrenoceptor population [22]. Of note, such differences are more noticeable in women than in men [22]. Second, these differences between VAT and SAT concern their secretory profile. More than a decade ago, AT has been also described as a secretory organ able to secrete bioactive molecules, the so-called adipokines, within the tissue as well as in the bloodstream. The first one to be described is leptin, which is involved in the lipostatic system leading to the hypothalamic control of energy expenditure and food intake [23]. The subcutaneous fat depot is the major source of leptin since higher secretion rates have been depicted than in VAT [24]. The other important adipokine is adiponectin (for review see [25]). The amount of SAT is correlated with adiponectin serum levels. Since adiponectin amount is inversely related with the settlement of type 2 diabetes, the "protective" role of SAT has been reinforced [26]. Proinflammatory cytokines as well as bioactive peptides, such as PAI-1 (Plasminogen Activation inhibitor-1) and Vascular Endothelial Growth Factor (VEGF), are more expressed in VAT and, for example, the amount of circulating VEGF is correlated with visceral obesity (respectively [27], [28], and [29]). Such metabolic/secretory differences between VAT and SAT are largely exacerbated with the increase of fat mass, since most of them are clearly linked to fat cell size. Indeed, adipocytes isolated from SAT are larger than those obtained from VAT and obesity-associated hypertrophied adipocyte emphasizes such a phenomenon and the physiological responses as well. The relative inertia of hypertrophied SAT adipocytes led to proposal for a role of SAT as a metabolic buffer via its long-term entrapment of excess FA and protection against ectopic fat deposition and lipotoxicity [30, 31].

Although the easiest way to describe WAT is to distinguish two main localizations with "site-specific properties," a more precise description can be made. For example, SAT is heterogeneous in terms of histological and metabolic features. One can discriminate the SAT very close to the skin and the deeper one that are separated by a fibrous membrane well known by the plastic surgeons [32]. Various characteristics of SAT from different regions from the body have also been described [33]. Concerning the so-called VAT major differences can be found depending on the part of the body since omental, mesenteric, abdominal, perigonadal, and retroperitoneal AT behave differently. For example, the blood flow from mesenteric and omental AT drains into the portal vein, while retroperitoneal and perigonadal does not [5]. Such differences are of major interest for the understanding of the role of different fat depots in the drainage of NEFA and its consequences in term of insulin resistance and liver dysfunction, but should also be considered when cancer progression is studied as well.

Several "nontypical" AT have been described such as the one included in the BM. It consists of smaller adipocytes loosely bound together when compared to the classical collagen-driven shape of those found elsewhere. BM mature adipocytes possess unique profiles and have been proposed to exhibit features of both WAT and BAT [34]. BM adipocytes might provide nutrients and secrete adipokines to support hematopoiesis when required [35]. However, the number of BM mature adipocytes correlates inversely with the hematopoietic activity of the marrow (seen for example during aging) and a very elegant study has recently demonstrated that adipocytes act as negative regulators of normal hematopoiesis [36]. Both in vitro and in vivo data suggest that adipocytes prevent hematopoietic progenitor expansion and differentiation while preserving the hematopoietic stem cell pool [36]. Other nontypical AT can be found throughout the body. For example, emerging evidence suggest that perivascular AT regulate vascular function through numerous mechanisms [37] and most of the lymph nodes are embedded in AT. Adipocytes around lymph nodes are equipped to amplify their capacity to respond to lymphoid cells. Lipolysis increases in surrounding adipocytes when a lymph node is activated during an immune challenge. Such modifications are found in the close environment of the lymph node suggesting a paracrine rather than systemic response (for review, see [38]). Such data reveal intradepot differences in terms of function as well as responsiveness leading to the hypothesis of "local" adipocytes specialized to supply adjacent tissues. In females, a specific AT is described as the mammary fat pad surrounding the mammary gland. It is involved in the gland formation and in the regulation of epithelial cell function and proliferation [39]. Intimate and bidirectional interaction with adjacent epithelium is one of the hallmarks of mammary adipocytes. However, these relationships are both stage and species dependent as we underlined in a recent review [40]. The secretory and metabolic profiles of these "nontypical" AT are poorly characterized. Significant differences could exist between them and SAT or VAT that could profoundly impact cancer/AT crosstalk.

These specificities in terms of depots as well as species raise the question of the source of adipose cells that should be used in experimental studies to depict the cancer/AT connection. The detailed description of all the available human and murine models can be found in the review written by Lafontan [7] that we recently extended to breast cancer studies [40]. In front of these data, oncologists interested by AT must be aware that special attention should be paid to the sources of fat cells used in coculture models as well as to the site of implantation of tumor cells when used in xenograft models in mice. Similarly, an extensive characterization of human AT adjacent to tumors will clearly contribute to decipher the importance of AT specificity in cancer progression. Obvious limitations to this concept are represented by the low number of available murine or human preadipocyte cell lines, able to differentiate in vitro in mature adipocytes, for such studies [7]. The use of primary cells, if it already appears optimal for adipocyte precursors, is very limited for mature adipocytes in view of their very short life span in culture [7]. Development of reliable 3D culture models certainly will represent a major technical breakthrough regarding this aspect.

7.3 Experimental Evidences That AT Play a Paracrine Role in Cancer Progression

7.3.1 Breast Cancer and Other Gynecological Malignancies

Breast Cancer

Breast cancer is the most frequently diagnosed cancer and the second most lethal cancer in women in developed countries. As breast is a fat-rich organ, questioning the relationship between epithelial cancer cells and the cellular components of the AT appears evident. At anatomical levels, the epithelial ducts and lobules are invested with a dense connective tissue matrix outside of which islands of AT occupy a large proportion of the stroma [41]. Therefore, invasive breast cancer after local invasion of the fibrous layer will get in direct cell–cell contact with the MAT. In addition, in in situ carcinomas, the secretion of mammary adipose cells might diffuse to the proximal tumor cells, explaining a potential involvement of AT in the early steps of carcinogenesis. Accordingly, the first data implicating AT in cancer progression in vivo [42] and in vitro [43] have been obtained using breast cancer models. The role of the various cellular actors of AT, as well as the molecular circuitry involved, begins to be more closely characterized. It is important to underline that, to date, the involvement of mature adipocytes, rather than adipose precursors, have been more convincingly demonstrated in human tumors. We have extensively reviewed recently the role of adipose cells in breast cancer progression. Therefore, the reader could refer to this review for a detailed description of the AT/breast cancer cells crosstalk [40].

What are the key results that characterize this crosstalk? The first element is that mature adipocytes are not inert to their surrounding and can be profoundly modified by tumor cells secretion. Consistent with the name already used for fibroblasts surrounding tumors (Cancer-Associated Fibroblasts, CAFs) or with macrophages of the tumor microenvironment (Tumor-Associated Macrophages, TAMs), we decided to name these cells Cancer-Associated Adipocytes (CAAs) [10, 44]. Found at

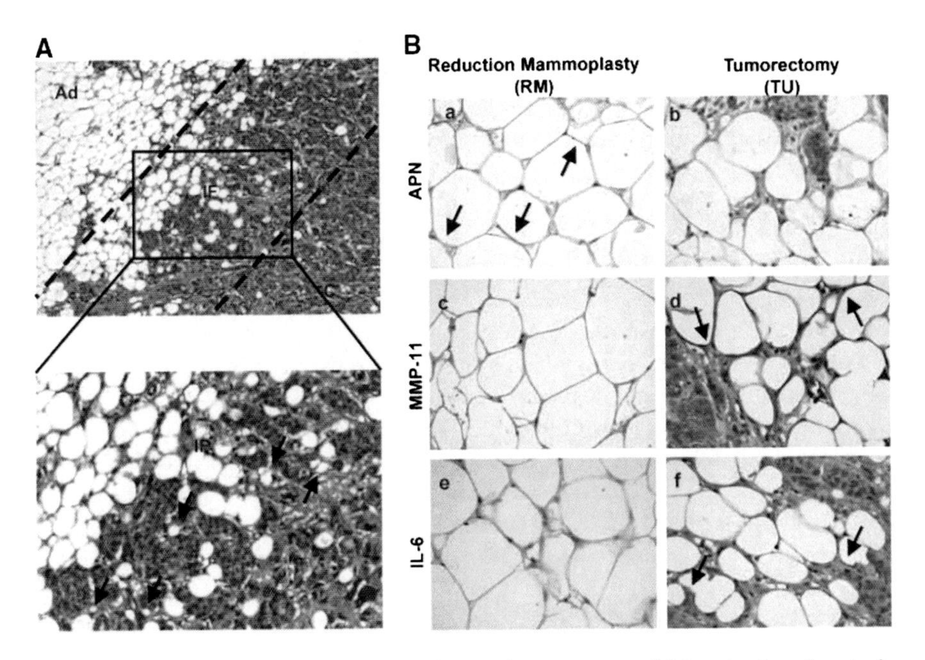

Fig. 7.1 Tumor-surrounding adipocytes in human breast tumor exhibit extensive phenotypic changes. (**A**) *Top*, histologic examination of an invasive breast tumor after H&E staining (original magnification ×200). *Ad* adipose tissue, *IF* invasive front, *C* tumor center. *Bottom*, histologic magnification of the invasive front of the tumor. Note that the number and the size of adipocytes were reduced (adipocytes of smaller size are indicated by *arrows*). (**B**) The expression Adiponectin (APN) **(a–b)**, MMP-11 **(c–d)**, IL-6 **(e–f)** was visualized (in *brown*) in adipocytes (lipids in *white*, nucleus in *blue*) located adjacent to invading cancer cells (*right columns*) as compared to normal adipose tissue (*left columns*). Reproduced with permissions from Dirat et al., Cancer Research 2011; 71: 2455–2465

invasive fronts of human breast tumors (see Fig. 7.1), the key features of CAAs are (1) decrease in size, delipidation, and loss of terminal differentiation markers, (2) overexpression of ECM and ECM-related molecules [11, 12], and (3) overexpression of inflammatory molecules such as IL-6 [10]. We have set up in our laboratory an original coculture system using breast cancer cells and either murine or human adipocytes (obtained from the in vitro differentiation of adipose progenitors present in the SVF of MAT) [10], the two cell populations being separated by an insert allowing the diffusion of soluble factors. The changes observed in adipocytes recapitulated the traits observed in human tumors including the overexpression of ECM-related and proinflammatory molecules highlighting the in vivo relevance of this coculture model [10]. The second element is that CAAs strongly stimulate tumor progression by enhancing tumor invasion both in vitro and in vivo. At least in vitro, this effect was observed in almost all the human and murine tumor cell lines tested independently of the expression of the estrogen receptor (ER), whereas the effect on cell proliferation was cell line dependent [10]. In addition to stimulation of tumor growth and invasiveness, our preliminary data demonstrate that CAAs might also promote a radioresistant phenotype [45], a result, if it is confirmed for other therapeutic strategies, could open important clinical perspectives.

As stated before, two major traits have been described to date that characterize the "activated state" of CAAs. The first one is an extensive ECM remodeling suggesting the involvement of ECM-signaling events in adipocyte/cancer crosstalk (for recent reviews see [46] and [47]). It has been shown that type VI collagen is involved in this process [11, 48]. Type VI collagen, while expressed by a number of other cell types, is abundantly produced and secreted by adipocytes. Collagen VI (COLVI) is composed of three chains, α1, α2, and α3, which associate to form higher-order complexes. Human breast cancer surrounding adipocytes exhibit increased COLVI α3 expression [48]. In vitro recombinant COLVI promotes GSK3β phosphorylation, β-catenin stabilization, and increases β-catenin activity in breast cancer cell lines, resulting in increased cell proliferation [48]. Accordingly, collagen (–/–) mice in the background of the mouse mammary tumor virus/polyoma virus middle T oncogene (MMTV-PyMT) mammary cancer model demonstrate dramatically reduced rates of early hyperplasia and primary tumor growth [11]. Interestingly, the COLVI form overexpressed in human and murine tumor is the carboxyterminal domain of collagen VIα3, a proteolytic product of the full-length molecule [11] that possesses stimulatory functions regarding tumor growth [11]. Therefore, different functional aspects of collagen VI overexpression might be envisioned with regard to tumor progression. First, the increase in matrix deposition might support tumor growth through direct mechanical constraints. Second, the increase in COLVI expression might contribute to the occurrence of the CAAs phenotype since it favors adipocyte metabolic dysfunction and inflammation [49]. Finally, it may acts as a signaling molecule most probably though cleavage fragments. This latter event could be dependent of the overexpression of metalloprotease 11 (MMP-11)/stromelysin 3. In fact, MMP-11 is overexpressed in tumor-surrounding adipocytes in vivo [12] and in vitro in our 2D coculture system [10]. Part of the MMP-11 effect could be relied on COLVI since MMP-11 is able to cleave the native α3 chain of collagen VI [50]. Dependently or not of COLVI, MMP-11 overexpression could also contributes to the "dedifferentiation" process observed in CAAs since it acts as a negative regulator of adipogenesis [12]. The second major event that takes place in CAAs is the occurrence of an inflammatory phenotype. We have shown that CAAs overexpress IL-6 both in vitro and in vivo and that, in the coculture system, IL-6 blocking antibodies (Abs) inhibits the CAAs-dependent proinvasive effect [10]. Very interestingly, human breast tumors of larger size and/or with lymph nodes involvement exhibit the highest levels of IL-6 in tumor-surrounding adipocytes [10]. As we see below, the increase in IL-6 in tumor-surrounding adipocytes has been observed in other tumors such as prostate cancer [51]. Concerning the role of two most extensively studied adipokines, leptin, and adiponectin, the reader is invited to refer to the previous chapter on the role of adipokines on cancer progression.

Taken together, these recent findings about the characterization of CAAs phenotype and function might have important clinical consequences. In fact, CAAs and adipocytes in obesity conditions share common traits including the excess in ECM deposition [52] and inflammation [6]. Therefore, it is tempting to speculate that the deleterious crosstalk between mature adipocytes and breast cancer is amplified in obesity conditions and contributes, at least for a part, to the poor prognosis of breast cancer observed in obese patients (for review see [40, 44]). In fact, excess bodyweight is not only associated to increased incidence of breast cancer in postmenopausal women but is also an independent negative prognosis factor independently of menopause status (for review see [40, 44]). Regarding a paracrine effect, this hypothesis implies, however, that these changes, fibrosis and inflammation, are found in MAT of obese patients. Recent data obtained in humans and animals suggest that obesity is associated in MAT with both adipocyte hypertrophy, macrophages infiltration, as well as IL-6 overexpression [53, 54], as we recently reviewed [40]. To our knowledge, increase matrix deposition in MAT of obese patients has never been studied. As these data concerning inflammation in MAT appears to partially validate the concept of an amplified detrimental paracrine crosstalk between cancer cells and adipocytes in obese conditions, experimental studies using 3D coculture and animal models in lean and obese conditions are eagerly awaited. In mouse models, preliminary results have been obtained that links obesity and the occurrence of high-grade adenocarcinomas, increased tumor growth, as well as increased angiogenesis and distant metastasis [55, 56]. These models will be very helpful to identify the paracrine mechanisms involved in the stimulation of tumor progression in order to set up specific strategies for the treatment of obese patients exhibiting aggressive diseases.

The AT also contains adipose progenitors, including ADSCs and preadipocytes, that are released after collagenase digestion of AT and short centrifugation and located in the so-called SVF. ADSCs display multipotency and proliferation capacities comparable to those of bone marrow MSC, while also having clear unique features (for a review see [16]). Elegant studies have recently showed that, in mice, ADSCs are recruited by experimental tumors and promote cancer progression [57, 58]. These studies using fluorescent-tagged cells show that these ADSCs contribute to form most vascular and fibrovascular stroma (pericytes, α-SMA$^+$ myofibroblasts, and endothelial cells) in murine tumors [58]. The contribution of these cells to tumor vascularization will be discussed in the next chapter. Concerning stromal (i.e., fibroblast-like) cells derived from ADSCs, coculture experiments using ADSCs and breast tumor cells show that tumor cells are able to "differentiate" ADSCs into CAF-like cells expressing α-smooth muscle actin (α-SMA), tenascin C, and fibronectin (for review see [40]). These data suggest that "activated" ADSCs could directly contribute to the dense network of fibroblasts and ECM (the so-called desmoplastic reaction) frequently observed in the center of breast tumors and, as such, be considered as a subpopulation of CAFs. Interestingly, this hypothesis was recently validated by a study demonstrating that these activated ADSCs led to ECM deposition and contraction in vivo thereby enhancing breast cancer tissue stiffness [59]. In turns, these activated ADSCs promote tumor growth, survival, and invasiveness through various signaling molecules including among others IL-6 [9] (for review see [40]). More recently, it has been demonstrated that ADSCs promote the expansion of cancer stem cells and induce EMT in the cancer cells in a PDGF-dependent manner [60]. Of note, most of these studies have been performed using

ADSCs from abdominal or subcutaneous sources with the noticeable exception of two studies that used MAT [9, 61]. As stated before since adipose depots have different biochemical profiles, it appears important to use breast-derived ADSCs when investigating the effects of "resident" stem cells on breast cancer cells. In addition, for technical reasons, it is difficult to prove in human tumors that a subpopulation of CAFs is indeed issued from activated ADSCs since for the moment they do not exhibit characteristic-specific traits that allow them to be distinguished among CAFs population. Nevertheless, studying their role in tumor progression is also highly relevant to the clinical situations as already underlined in the introduction section.

The clinical implication of this deleterious crosstalk between both mature and progenitor adipose cells might also have impact on the oncological safety of fat grafting. In plastic and reconstructive surgery, autologous fat grating enables soft tissue augmentation and is increasingly used not only for cosmetic indications but also for correction of defect following breast cancer treatment. Accordingly, one should be aware that transfer of "activated" AT (activation being due to hypoxic and mechanical stress due to the removal of the AT before reinjection) might theoretically favor relapse of breast cancer. According to one controlled clinical study performed recently, this procedure appears to be relatively safe with the exception of in situ carcinomas that need to be confirmed [62], for review see [40] and [63]. Recent fat transfer and lipoinjection protocols have focused on the addition of autologous ADSCs or freshly isolated adipose stromal vascular cells to promote retention of graft volume [64]. This type of procedures should be considered with even greater caution with respect to conventional technique. In fact, a recent study that will be discussed extensively in the next chapter by Bertolini has shown that the CD34+ fraction of ADSCs present in the lipofilling material contributed to tumor vascularization and significantly increased tumor growth and metastases in several orthotopic models of human breast cancer [65]. In conclusion, AT has emerged these last 5 years as an integral and active component of breast cancer microenvironment. It is likely that the continued characterization of these interactions, and the molecular identification of key mediators, will provide new insights into tumor biology and suggest further novel therapeutic options.

Ovarian and Endometrial Cancers

Ovarian cancer is a highly fatal disease, with only about 40 % of women with ovarian cancer still alive more than 5 years postdiagnosis. This poor survival is largely attributable to the fact that a majority of ovarian cancer in developed countries is diagnosed with metastatic spread. The omentum, a peritoneal organ rich in AT and immune cells, has been shown to be a preferred site of metastatic dissemination in ovarian cancer patients. Omental dissemination, which is often accompanied by ascites, facilitates the further spread of the tumors [66]. In a recent study, Nieman et al. demonstrate that ovarian cancer cells preferentially migrate towards omental adipocytes-conditioned medium rather than to adipocytes from other sources in a

CXCR-1/IL-8 axis-dependent manner [13]. Some others cellular components of the omentum, such as fibroblasts [67] or immune cells [68], have also been implicated in peritoneal dissemination. Accordingly, additional studies are clearly needed to determinate, which cellular components of the omentum predominate to explain the omental "tropism migration" of ovarian cancer cells. Once located to the omentum, the tumor cells must found optimal conditions to survive and proliferate. A study published last year directly implicated mature adipocytes in this process by first showing that omental adipocytes favored the growth of ovarian cancer cells both in vitro and in vivo [13]. The mechanism involved in this crosstalk is rather original. Indeed, in this work, the authors have shown that ovarian cancer cells induce lipolysis in mature adipocytes that is followed by a transfer of FFAs between omental adipocytes and cancer cells as shown both in experimental models and in human tumors [13]. The fact that coculture of adipocytes with the tumor cells stimulate the mitochondrial β-oxidation activity in the tumor cells suggest, without demonstrating it directly, that the omental adipocytes might serve as an energetic source (i.e., NEFAs delivery) to support tumor growth. However, other alternate roles for these NEFAs might be envisioned including the production of lipid signaling molecules as well as their conversion into specific lipid species for incorporation into membranes. One of the major interests of this work concerns the identification of a pharmacological target, the Fatty Acid Binding Protein-4 (FABP-4), whose inhibition could prevent the deleterious dialogue between adipocytes and cancer cells [13]. FABP4 is the major FABP expressed in adipocytes, a family of small, highly conserved, cytoplasmic proteins that bind long-chain fatty acids and other hydrophobic ligands. It is thought that FABPs contribute to fatty acid uptake, transport, and metabolism (for review see [69]). Although mainly expressed by adipocytes, FABP4 was also expressed at lower levels by ovarian cancer cells and its expression in tumor was upregulated in coculture in vitro as well as at the interface between cancer cells and adipocytes in vivo [13]. Lipid content was decreased in tumor cells cocultivated with *Fabp4*$^{-/-}$ adipocytes and in an orthotopic model of ovarian cancers, very few peritoneal metastases were detected in *Fabp4*$^{-/-}$ mice in opposition to wild-type mice. Interestingly, pharmacological inhibitors of FABP4 already exist. Use of these inhibitors decrease adipocyte-cocultivated tumor cells invasion in vitro as well as tumor growth in vivo [13]. Taken together, these results demonstrate that FABP4 behave as a key mediator of ovarian cancer cell–adipocyte interactions in the host and potentially in the cancer cells by increasing lipid availability and supporting metastasis. These findings have important conceptual and pharmacological consequences. They suggest that a metabolic crosstalk is established between adipocytes and tumor cells, adipocytes-derived NEFAs being used by cancer cells to fuel tumor progression. These results reinforced the new concept of a metabolic coupling between tumor and stroma cells, also demonstrated for CAFs, the stroma being catabolic (lipolysis, glycolysis, etc.) whether the cancer cells are anabolic [70]. Whether or not this crosstalk exists for other tumor types remain to be determined. In vitro, a transfer of lipids is observed when omental adipocytes are cocultivated with breast and colon cancer [13]. However, as stated above, susceptibility to lipolysis is different among the AT depots and should be confirmed for breast cancers using appropriate adipocytes [22]. Also, in favor of a general phenomenon, it is interesting to note that lipid translocation has been demonstrated between prostate cancer cells and adipocytes using Fourier transform infrared (FTIR) microspectroscopy [71]. Finally, these findings offer new therapeutics opportunities to treat ovarian cancer. Uncovering enzymes that could be crucial in the regulation of this process in tumor cells (from molecules able to induce lipolysis in adipocytes to molecules involved in NEFAs uptake, processing, or oxidation) would pave the way to the development of new pharmacological approaches to target tumor cell metabolism.

Few data exist concerning the role of others components of AT in ovarian cancer progression. As for breast cancer, ovarian cells are able to "transdifferentiate" ADSCs towards a CAF-like phenotype with overexpression of α-SMA, both TGF-β [72], and lysophosphatidic acid (LPA) having been involved in this process [73]. One can notice that LPA can be largely released by mature adipocytes [74]. Another study performed in mice has recently demonstrated that ovarian cancer cells recruit the nearby ADSCs to form most vascular and fibrovascular stroma (pericytes, α-SMA$^+$ myofibroblasts, and endothelial cells) [58]. These results are in agreement with the fact that in vitro, CAF-like cells derived from activated ADSCs exhibit increased expression of proangiogenic molecules including VEGF and SDF-1 [75]. Again, as we have already underlined in breast cancer, further validation in human tumors are needed to fully validate the involvement of ADSCs in ovarian cancer progression. What do all these experimental results tell us about the link between ovarian cancer and obesity? At the present time, studies that have examined the association between obesity and ovarian cancer survival have provided conflicting results [76]. A recent meta-analysis concluded that obese women with ovarian cancer appear to have slightly worse survival compared to nonobese women [76]. According to the important role that appears to play omental adipose tissue in the progression of the disease, anthropometric measures other than increased BMI, for example waist-to-hip ratio or waist circumference, might be better measures of adiposity in terms of ovarian cancer progression risks.

The last gynecological malignancy that might be related to paracrine effects of AT is endometrial cancer. As seen in previous chapter, the occurrence of endometrial cancer is strongly linked to obesity [77]. The relationship between endometrial cancers and obesity has generally been attributed to "uncompensated" estrogens and others adipokines secreted by AT acting in an endocrine manner. Are there arguments suggesting that AT can act in a paracrine manner? To our knowledge, no data has been published to date on the paracrine effect of mature adipocytes on endometrial cancer progression. Concerning ADSCs, it has been recently demonstrated that adipose precursors isolated from the omental AT specifically stimulate in vitro and in vivo the growth of the endometrial cancer cell line, Hecal, as well as the tumor vascularization in vivo [78]. Very interestingly, this effect was smoldered when ADSCs from SAT were used. These results strongly suggest that ADSCs of the omentum have specific functional characteristics that make them more prone to stimulate tumor progression of neighboring endometrial cancers [78].

7.3.2 *Prostate Cancer and Other Cancers from the Genitourinary Tracts*

Prostate cancer is the most common malignancy in males in Western countries, representing the second leading cause of cancer death. Prostate is surrounded by AT and tumor admixed with periprostatic fat is the most easily recognized manifestation of extraprostatic extension, a well-established adverse prognostic factor for prostate cancer [79, 80]. Periprostatic AT (PPAT) is considered as VAT, but the specificities of this depot in terms of metabolism and adipokines secretion remain largely unknown. At laboratory levels, the contribution of this tissue to cancer progression has been first suggested by the report of Finley et al. that analyzed the PPAT features in patients undergoing prostatectomy for cancer [51]. In this study, the authors found that the level of IL-6 secreted by PPAT-conditioned medium (CM) was almost 375 times greater than the circulating levels of the cytokine in the same patient. Both IL-6 levels in PPAT and activation of IL-6 related signaling pathways were correlated to tumor aggressiveness [51]. Therefore, this study strongly suggests that PPAT represents an important source of IL-6 that favors tumor progression. Interestingly, several studies already reported that increased serum IL-6 and soluble interleukin-6 receptor levels are associated with aggressiveness of the disease and with a poor prognosis in prostate cancer patients, underlying the importance of this pathway in PC progression (for review see [81]). The nature of the cells that contribute to secrete IL-6 in the PPAT of patients exhibiting aggressive tumors remains undetermined since both mature adipocytes, ADSCs as well as inflammatory cells potentially present within the AT have been described as IL-6 secreting cells [6]. However, the low number of inflammatory cells in the periprostatic fat specimens led the authors to propose that this secretion is in fact dependent of adipose cells [51]. Recent studies suggest that, like in breast cancer, a bidirectional crosstalk exists between PC cells and surrounding AT. Explants of PPAT incubated with CM of the aggressive prostate cancer cell line PC3 exhibit increased expression in adipokines associated with cancer progression (osteopontin, TNFα, and IL-6) [82]. Conversely, CM obtained from explants of PPAT stimulates the proliferative and migratory capacities of PC3 cells [83].

Two additional preliminary results from these studies need attention. First, the SVF fraction appears to play little role in this crosstalk since the expression of inflammatory cytokines of SVF cells was unaffected by the CM of PC3 cells [82] and the CM obtained from the SVF fraction alone do not simulated the PC3 survival and migratory capacities [83]. These results need to be confirmed as another study has recently demonstrated that ADSCs promote prostate tumor growth in coinjection experiments in mice [84]. Second, explants cultures using AT of periprostatic origin are most effective in promoting proliferation and migration of PC-3 cells than AT of other visceral origin suggesting an exquisite depots specificities [83]. Taken together, these first compelling results suggest that PPAT plays a role in prostate cancer progression by contributing to local inflammation through IL-6 secretion and that the crosstalk is established with mature adipocytes rather than adipose cells precursors. Although preliminary, these results also suggest that PPAT exhibit specific traits that should be taken in account in forthcoming studies. Finally, it is interesting to note that the crosstalk of PC cells with surrounding adipose tissue exhibit similarities with those observed in breast cancer cells, observations that might lead to common therapeutic strategies. These results are also in accordance with recent clinical findings showing that whether obesity is not a consistent risk factor for prostate cancer incidence, a high BMI is associated with higher-grade tumors and nonorgan-confined disease [85]. Interestingly, the association between visceral obesity and worse prognosis appears stronger when the importance of peri-prostatic fat, rather than BMI, is considered [86]. In light of the increasing world-wide incidence of obesity, the identification of obesity as a risk factor for aggressive prostate cancer is important because it may be one of the few modifiable risk factors for this disease. Obesity is also associated to a higher risk of developing renal cell carcinomas (RCC) [87]. Interestingly, in opposition to the observations made for other cancers, epidemiological evidences suggest that obesity may confer an improved tumor-specific survival for localized RCC postnephrectomy [88]. To our knowledge, the paracrine role of adipose cells in renal cell carcinomas occurrence and progression has not been investigated.

7.3.3 Gastrointestinal Cancers

Gastrointestinal tract malignancies are very common cancers that occur within the digestive system organs—the stomach, liver, pancreas, colon, and rectum. As VAT is at proximity of the gastrointestinal tract, a similar crosstalk as the one described above for prostate or breast cancer is likely to take place and to influence digestive cancer development and progression. In addition, as ovarian cancers, malignant cells from intra-abdominal primary tumors including colon, stomach, and pancreas often metastasize into the AT of the peritoneal cavity, an event associated to a very poor prognosis for the patients. Of note, the lipid transfer demonstrated between omental AT and ovarian cancers appears also to exist with colon and gastric cancer cell lines [13]. Whereas the endocrine effect of adipokines on digestive tumors has been largely reported [89], few studies have concerned the paracrine effects of AT. Data published to date are reviewed here according to the principal tumor subtypes that have been studied regarding the paracrine effect of AT.

Colon Cancer

Colon cancers are at close anatomical proximity of VAT. Cancer cells after invasion of the basement membrane will get in direct cell–cell contact with the VAT. In addition, in noninvasive carcinomas, the secretion of adipose cells might diffuse to the proximal tumor, explaining a potential role of AT in the early steps of carcinogenesis. As for breast cancer, clinical recent evidences demonstrate that colon cancers modify the surrounding AT. A significant reduction in both lipoprotein lipase (LPL) and fatty acid synthase (FAS) gene expression and activity levels is observed in AT adjacent to colon cancer compared to those detected in paired AT distant from neoplasia [90]. This is associated to a tumor-induced impairment in the formation and lipid-storing capacity of tumor surrounding AT [90]. Again, these results probably reflect the relative dedifferentiation observed in CAAs. Interestingly, decrease in LPL and FAS is also observed in adipocytes cocultivated with breast cancer in our 2D coculture system (Wang et al., unpublished results). This crosstalk has been partially depicted using in vitro 3D coculture system between human cancer cell lines and primary murine adipocytes. This study demonstrates that both isolated mature adipocytes and preadipocytes have proliferative effects on colon cancer cell lines, although the effect of preadipocytes is more potent [91]. This trophic effect of mature adipocytes on the colon cancer cell lines was no longer observed when mature adipocytes from leptin-deficient mice were used (ob/ob mice) [91]. The effect of leptin on colon cancer cell lines has been confirmed by other studies. In vitro, exposure to leptin enhances the growth of colon cancer cell lines [92] and also enhances motility and invasiveness of such cells [93, 94]. Colon cancer stem cells respond to leptin with increased cell proliferation, enhanced growth in soft agar, and improved sphere formation [95]. In addition, leptin counteract the cytotoxic effects of 5-fluorouracil, a common colon cancer therapeutic agent [95]. This interference of adipokines with response to chemotherapy would be to our point of view an important aspect that needs to be taken in account in future studies. However, all these studies performed with recombinant leptin do not assess that in vivo the effect of this hormone would be rather of paracrine than endocrine nature. As for other cancers, it is tempting to speculate that the negative crosstalk between AT and colon cancer is amplified in obesity conditions. In fact, several recent meta-analyses have correlated obesity with an increased incidence of colon cancers [96–98]. This correlation between obesity and the occurrence colon cancer has been confirmed in mouse models [99, 100]. Obesity has also been associated with poorer prognosis in some studies, but this evidence is still debated (for review see [101]). In vitro studies have addressed the question of tumor aggressiveness in response to adipocyte-conditioned medium (Ad-CM) obtained from VAT of obese patients. Interestingly, a higher increase in cell proliferation of colon cancer cell lines was observed in the presence of Ad-CM from VAT of centrally obese patients as compared to lean patients. Moreover, this effect was inhibited in the presence of neutralizing VEGF antibodies [102]. To conclude, these early works linking AT to the development and aggressiveness of colon cancers, although scant, compared for example to breast cancer, are very promising and should be consolidated.

Gastric Cancer

A paracrine dialogue between cancer cells and adipose tissue is possible if one considers that gastric submucosa and subserosa might contain AT. Most of studies performed to date have implicated ADSCs rather than mature adipocytes. As for other cancers, a bidirectional crosstalk is established between cancer cells and adipose progenitors therefore contributing to tumor progression. Using a 3D coculture system, Nomoto-Kojima et al. have demonstrated that ADSCs promote the growth/invasion and suppress the apoptosis of the cancer cells. Notably, ADSCs cause the cancer cells to inhibit the expression of HER2, which is a key molecule for the molecular-targeted therapy of advanced gastric cancer. In turn, gastric cancer cell types induce a CAFs phenotype in ADSCs including expression of α-SMA [103]. A recent study showed that gastric tumors and cell lines express the chemokine receptor CXCR4 that has been previously described as a strong determinant in breast cancer cell dissemination [104]. As ADSCs express one of the CXCR4 ligand, CXCL-12 or SDF-1, the authors show that adipose progenitors promote cell growth, migration, and invasion depending on a CXCL12/CXCR4 axis [105]. However, this study was performed using ADSCs isolated from SAT. Similar studies should be performed using omental AT to see if the CXCR4/CXCL-12 axis might favor the dissemination of gastric cancer cells to the peritoneum. Strikingly, no significant association has been found between obesity or waist circumference and increased incidence of gastric cancer [98]. Nevertheless, as for prostate cancer, significant trends of increasing risk with higher BMI index values were observed for death from stomach cancer in men [106]. This certainly deserves further investigation.

Pancreatic Cancers

During the last decade, pancreatic cancer has become the fourth leading cause of cancer-related death in the USA and the sixth leading cause in Europe. Despite major advances in surgical techniques and adjuvant therapies, overall 5-year survival remains under 5 %. While very few, if any, laboratory studies have been performed to date on the crosstalk between pancreatic cancers and AT, several clinical data have suggested that an adipose-rich environment leads to a deleterious outcome on this disease. First, it has been demonstrated that peripancreatic fat invasion is correlated to a poorer survival for pancreatic cancers [107]. Recent epidemiologic studies also suggest that obesity doubles the relative risk of pancreatic cancer [98]. In addition, central adiposity has been shown to be an independent risk factor in development of pancreatic cancer as well as to contribute to a poorer survival [108]. Interestingly, it has been demonstrated that increased pancreatic fat (pancreatic steatosis) promotes dissemination and lethality of pancreatic cancer [109]. Mice models have corroborated these data, including the role of intrapancreatic fat [110], and underlined the deleterious effect of obesity on pancreatic cancer aggressiveness [111, 112]. Concerning ADSCs, a study demonstrated that these cells strongly inhibit pancreatic cancer proliferation, both in vitro and in vivo and induce tumor cell death by altering cell cycle progression, in apparent opposition with all the findings obtained with other cancer types [113]. Given the importance of obesity in the evolution of pancreatic cancer, this model certainly represents a very interesting target for fundamental studies, reinforced by the need of new therapeutic approaches in this highly fatal disease.

7.3.4 Melanomas

Among the different types of skin cancer, no clear link between AT, obesity, and nonmelanoma skin cancer (basocellular and spinocellular carcinomas) has been established by contrast to melanoma. Melanoma, that results from melanocytes transformation, is a highly malignant tumor type, which is characterized by its tendency to give rise to metastases [114]. The number of cases of melanoma worldwide is increasing more rapidly than any other type of cancer. Indeed, the incidence of melanoma has more than tripled in the white population over the last 20 years [115]. As for other types of cancers, melanoma is not only composed by the malignant cells but also by the supporting stroma, which includes fibroblasts, endothelial cells, immune cells, soluble molecules, and the extracellular matrix (ECM) [116]. Melanoma can be cured only if detected and treated very early, before it begins to disseminate. Superficial forms of melanoma spread out within the epidermis, whereas when tumor cells have grown through the basement membrane and reach the deeper layer of the skin (vertical growth phase) it is considered as invasive [117]. Little attention has focused on AT although it is the main component of the deeper layers of the skin, the hypodermis. Therefore, the bioactive molecules secreted by AT may contribute to initiate the vertical growth phase of primary melanomas and to support the growth of locally invasive tumors. Interestingly, higher adipocytes content in melanoma tumors correlate with increased melanoma metastatic potential and decreased survival rates of patients [118]. At laboratory levels, there is evidence that a crosstalk is established between melanoma cells and AT and that the latter contribute to tumor aggressiveness. Using a syngeneic mouse melanoma model, Wagner et al. recently demonstrated that peritumoral AT exhibited reduced adipocyte size, extensive fibrosis, increased angiogenesis, and a dense macrophage infiltrate [119]. The observed phenotype observed for mature adipocytes is clearly reminiscent of the one previously described for breast cancers [10]. Of note, it has also been demonstrated that melanoma conditioned medium is able to induce lipolysis in mature adipocytes [120] as described for ovarian cancers [13]. However, in the study performed by Wagner et al. [119], the strong inflammatory response observed in peritumoral AT is mainly due to macrophages infiltration. These results suggest that even if mature adipocytes are the primary target of the established crosstalk with tumor cells, they act by recruiting accessory inflammatory cells [119]. Interestingly, tumors implanted at a site distant from subcutaneous, anterior adipose tissue were strongly growth delayed, had fewer blood vessels, and were less populated by CD11b (+) macrophages, supporting the idea that AT fuels the growth of malignant cells. In addition, a study suggests that mature adipocytes might be an important direct component of the AT/melanoma crosstalk. In fact, conditioned medium from mature adipocytes (obtained from the in vitro differentiation of the 3T3-L1 preadipocyte cell line) strongly stimulates the migratory and invasive abilities of melanoma cells in association with increased expression of EMT genes [121]. Epidemiological studies have suggested that obesity increases the incidence of melanoma development [122]. Nevertheless, it is not clear if this correlation is restricted or not to men [98]. Concerning carcinogenesis, a study demonstrated that obese mice exhibit greater susceptibility to chronic UV-B induced inflammation as well as decrease UV-induced apoptosis linking for the first time UV response and metabolic state [123]. In mice, there is also convincing data showing that obesity dramatically increases the growth as well as metastasis ability of established tumors (B16BL6 cell line) using both genetic and high-fat diet (HFD) obesity models [124–126]. Although the underlying mechanisms are poorly understood, it is interesting to note that in one study [125], a significant increase in caveolin-1 (Cav-1) and neutral fatty-acid synthase (FASN) expression was observed in the tumors obtained from obese as compared to lean mice. Inhibition of Cav-1 or FASN expression decreases the aggressiveness of the tumors in obese animals [125]. Implication of VEGF-related pathways with increased of both VEGF contents as well as VEGF-R in tumors xenografted in obese mice as compared to lean one has also been proposed [124]. One may not exclude that obesity affects melanoma progression predominantly through endocrine mechanisms since for example serum from obese mice reproduces the increase in metastatic ability as well as the EMT previously observed with 3T3L1-CM [121]. Concerning melanoma, the link observed between obesity and tumor aggressiveness is rather convincing in mouse models highlighting the need to consider the AT as an integral component of melanoma stroma in further studies.

7.3.5 Hematological Malignancies

The BM consists of developing hematopoietic cells and nonhematopoietic cells, the latter collectively termed the BM microenvironment. Major nonhematopoietic cell types that form the BM microenvironment include mesenchymal stem cells (MSCs) and mesenchymal-derived cells (including adipose cells), endothelial cells, and cells of the sympathetic nervous system. Recent data has indicated that cells of the BM microenvironment, including adipocytes, may also contribute to hematopoietic diseases (for review see [127]). As seen in the Sect. 7.2 describing the specificity of AT, BM adipocytes appear primarily to negatively affect hematopoiesis. What are the consequences of this crosstalk established under physiological conditions on hematological diseases? Few reports have been published to date, but they suggest on the opposite that mature adipocytes stimulate leukemic cells proliferation and survival. For example, leukemic cells of some patients with acute myeloblastic leukemia (AML), acute lymphoblastic leukemia (ALL), and chronic myeloid leukemia (CML) express the leptin receptor. Leptin has stimulant effects on proliferation of leukemia cells as well as antiapoptotic effects (for review see [128]). A paracrine role in this effect is likely since BM adipocytes have been shown to be a significant source of leptin [129]. More conclusive results have been obtained in specific leukemia subtypes. In ALL, a study recently highlights the role of the adipocyte in fostering leukemia chemotherapy resistance both in vitro (2D coculture system) and in vivo [130]. Interestingly, adipocytes prevented chemotherapy-induced apoptosis by upregulating prosurvival signals [130]. From a clinical point of view, it is interesting to note that obesity in children is associated with an increased recurrence of ALL in intermediate and high-risk patients [131], while by contrast no clear link between obesity state and AML prognosis has been established to date [132, 133]. In multiple myeloma (MM), neighboring adipocytes also participate to the disease by affecting tumor cell proliferation, apoptosis, and migration [134]. Thus, in contrast to solid tumors, few studies are clearly addressed the role of adipocytes in hematological diseases despite their emerging role in hematopoiesis. Certainly, future studies should address their role in leukemic stem cell survival and response to treatment among the expected answers.

7.4 Conclusions

The relationship between AT and cancer is complex and involves both paracrine and endocrine effects whose relative contribution to tumor progression remains to be determined. Regarding paracrine effects, we have underlined in this chapter the need to consider the appropriate neighboring AT for each cancer subtypes in experimental studies. In fact, there is clear variations between the different AT in terms of secretion and sensitivity to lipolysis for example [22] and depots specificities start also to be uncover regarding their abilities to stimulate tumor progression. This is the case for example of omental adipocytes that are more able than other adipocytes to chemoattract ovarian cancer cells [13]. This is also the case for periprostatic AT that exhibit higher capacity than other VAT to simulate prostate cancer proliferation and migration [83]. Nevertheless, regarding AT/cancer crosstalk, there are common features found in several cancer subtypes. First, it is very important to underline that adipose cells are not inert to their surrounding and that their phenotype are profoundly modified by cancer cell secretions. Mature adipocytes acquire a CAAs phenotype observed in vivo in human breast [10, 12] and colon cancers [90] as well as in xenografts models of melanoma [119]. Regarding ADSCs, they are activated in vitro and in animal models by tumor secretion and are likely to contribute, in addition to tumor vascularization, to a subpopulation of CAFs present in the tumor stroma of several cancer including breast and ovarian cancers [57, 58] (for review see [40] and [63]). Thus, working with conditioned medium of naïve or unprimed adipocytes might introduce experimental bias since the impact of tumors will not be taken in account. Second, one of the principal hallmarks of the adipose tissue/cancer crosstalk appears to imply the invasive properties of tumor cells as found in obese patients that frequently exhibit an increase in local and distant metastasis convincingly demonstrated for breast and prostate cancers [40, 44, 85]. Two mains pathways have been described to date that could contribute to this deleterious effect and that represent new strategic avenues to be explored in patients, especially in obesity conditions. The first one is inflammation since both mature and precursor adipose cells have been shown to contribute to tumor progression through for example IL-6 secretion [9, 10]. Of note, accumulation of macrophages and lymphocytes in the AT

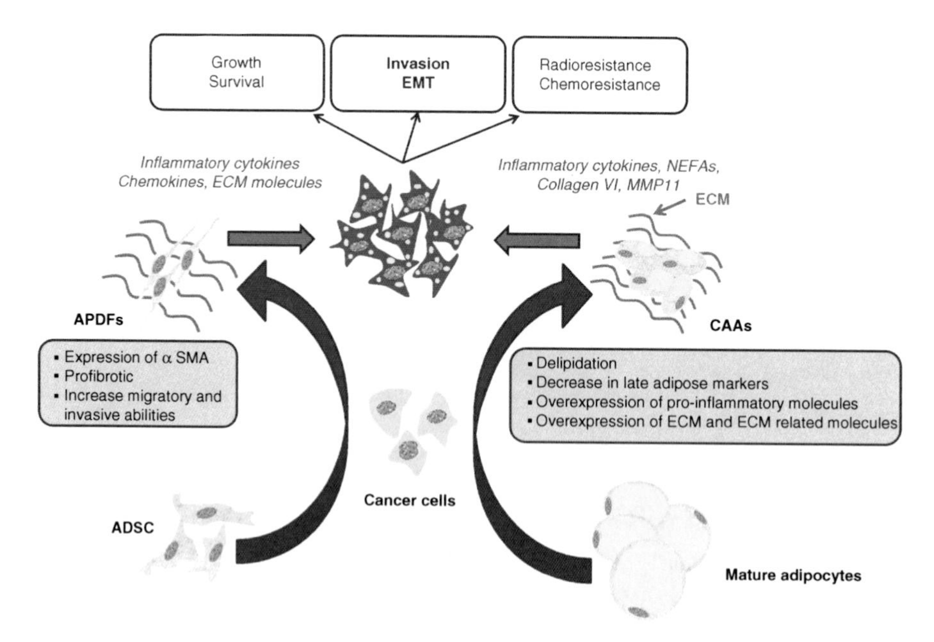

Fig. 7.2 Paracrine crosstalk between tumor cells and adipose cells contribute to cancer progression. Cancer cells could locally activate via soluble growth factors and cytokines, mature adipocytes, and adipose progenitor cells. Mature adipocytes exhibit a CAAs phenotype, while activated progenitors or Adipose-Progenitors Derived Fibroblasts (ADPFs) exhibit similar characteristics than CAFs. In turns, both ADPFs and CAAs clearly contribute to increase the aggressiveness of tumor cells by increasing tumor growth, tumor invasiveness, and resistance to antitumor therapy including ionizing radiation. This deleterious crosstalk might be amplified in obese patients contributing to the poor prognosis of cancers frequently observed in this subset of patients

of obese patients could also contribute to this detrimental inflammatory state. The second is the metabolic crosstalk that seems to be established between the two compartments. Transfer of NEFAs exists between mature adipocytes and tumor cells and this transfer supports tumor progression [13, 71]. Identifying the key molecular events involved in this process and more especially the fate of NEFAs within the cell will pave the future for developing new antitumor drug approaches. Third, one of the effect of AT that should be investigated is its role in the resistance to therapy. We have demonstrated that mature adipocytes contribute to radioresistance in breast cancers [45]. Others studies highlighted the importance of leptin in resistance to 5-fluorouracil, a common colon cancer therapeutic agent [95], the role of mature adipocytes in ALL sensitivity to Vinca alkaloids in vitro and in vivo [130] and the potential role of ADSCs in resistance to HER2 therapy in gastric carcinomas [103]. To our point of view, this aspect needs to be explored carefully since obesity is commonly associated to poor prognosis, which could also be caused to increased resistance to therapy. All these main results are summarized in Fig. 7.2. Since population of obese people is constantly increasing, it is of fundamental and of clinical interest to further study the relationship existing between adipose and cancer cells in order to prevent and treat this subset of aggressive diseases. Taken together, these data suggest that clinicians should consider obesity and metabolic factors, not only in assessing cancer risk but also in the design of clinical trials in oncology.

Acknowledgments Studies performed in our laboratories were supported by the French National Cancer Institute (INCA PL 2006–035 and INCA PL 2010–214 to PV and CM), the "Ligue Régionale contre le Cancer" (Comité du Lot, de la Haute-Garonne et du Gers to CM), the Fondation de France (to PV and CM), the Association pour la Recherche sur les Tumeurs Prostatiques ARTP (to CM), and the University of Toulouse (AO CS 2009 to CM). We would like to thank Dr Max Lafontan for critical reading of the manuscript.

References

1. Mueller MM, Fusenig NE. Friends or foes—bipolar effects of the tumour stroma in cancer. Nat Rev Cancer. 2004;4(11):839–49.
2. Schafer M, Werner S. Cancer as an overhealing wound: an old hypothesis revisited. Nat Rev Mol Cell Biol. 2008;9(8):628–38.
3. Kalluri R, Zeisberg M. Fibroblasts in cancer. Nat Rev Cancer. 2006;6(5):392–401.
4. Mantovani A et al. Cancer-promoting tumor-associated macrophages: new vistas and open questions. Eur J Immunol. 2011;41(9):2522–5.
5. Bjorndal B et al. Different adipose depots: their role in the development of metabolic syndrome and mitochondrial response to hypolipidemic agents. J Obes. 2011;2011:490650.
6. Ouchi N et al. Adipokines in inflammation and metabolic disease. Nat Rev Immunol. 2011;11(2):85–97.
7. Lafontan M. Historical perspectives in fat cell biology: the fat cell as a model for the investigation of hormonal and metabolic pathways. Am J Physiol Cell Physiol. 2012;302(2): C327–59.
8. Halberg N, Wernstedt-Asterholm I, Scherer PE. The adipocyte as an endocrine cell. Endocrinol Metab Clin North Am. 2008;37(3):753–68. x–xi.
9. Walter M et al. Interleukin 6 secreted from adipose stromal cells promotes migration and invasion of breast cancer cells. Oncogene. 2009;28(30):2745–55.
10. Dirat B et al. Cancer-associated adipocytes exhibit an activated phenotype and contribute to breast cancer invasion. Cancer Res. 2011;71(7):2455–65.
11. Iyengar P et al. Adipocyte-derived collagen VI affects early mammary tumor progression in vivo, demonstrating a critical interaction in the tumor/stroma microenvironment. J Clin Invest. 2005;115(5):1163–76.
12. Andarawewa KL et al. Stromelysin-3 is a potent negative regulator of adipogenesis participating to cancer cell-adipocyte interaction/crosstalk at the tumor invasive front. Cancer Res. 2005;65(23):10862–71.
13. Nieman KM et al. Adipocytes promote ovarian cancer metastasis and provide energy for rapid tumor growth. Nat Med. 2011;17(11):1498–503.
14. Calle EE, Kaaks R. Overweight, obesity and cancer: epidemiological evidence and proposed mechanisms. Nat Rev Cancer. 2004;4(8):579–91.
15. Gir P et al. Human adipose stem cells: current clinical applications. Plast Reconstr Surg. 2012;129(6):1277–90.
16. Zhang Y, Bellows CF, Kolonin MG. Adipose tissue-derived progenitor cells and cancer. World J Stem Cells. 2010;2(5):103–13.
17. Nedergaard J, Bengtsson T, Cannon B. Unexpected evidence for active brown adipose tissue in adult humans. Am J Physiol Endocrinol Metab. 2007;293(2):E444–52.

18. Kozak LP, Koza RA, Anunciado-Koza R. Brown fat thermogenesis and body weight regulation in mice: relevance to humans. Int J Obes (Lond). 2010;34 Suppl 1:S23–7.
19. Despres JP, Lemieux I. Abdominal obesity and metabolic syndrome. Nature. 2006; 444(7121):881–7.
20. Hotamisligil GS. Inflammation and metabolic disorders. Nature. 2006;444(7121):860–7.
21. Zierath JR et al. Regional difference in insulin inhibition of non-esterified fatty acid release from human adipocytes: relation to insulin receptor phosphorylation and intracellular signalling through the insulin receptor substrate-1 pathway. Diabetologia. 1998;41(11):1343–54.
22. Lafontan M, Langin D. Lipolysis and lipid mobilization in human adipose tissue. Prog Lipid Res. 2009;48(5):275–97.
23. Zhang Y et al. Positional cloning of the mouse obese gene and its human homologue. Nature. 1994;372(6505):425–32.
24. Van Harmelen V et al. Leptin secretion from subcutaneous and visceral adipose tissue in women. Diabetes. 1998;47(6):913–7.
25. Kadowaki T et al. Adiponectin and adiponectin receptors in insulin resistance, diabetes, and the metabolic syndrome. J Clin Invest. 2006;116(7):1784–92.
26. Drolet R et al. Fat depot-specific impact of visceral obesity on adipocyte adiponectin release in women. Obesity (Silver Spring). 2009;17(3):424–30.
27. Fontana L et al. Visceral fat adipokine secretion is associated with systemic inflammation in obese humans. Diabetes. 2007;56(4):1010–3.
28. Fain JN et al. Comparison of the release of adipokines by adipose tissue, adipose tissue matrix, and adipocytes from visceral and subcutaneous abdominal adipose tissues of obese humans. Endocrinology. 2004;145(5):2273–82.
29. Ledoux S et al. Angiogenesis associated with visceral and subcutaneous adipose tissue in severe human obesity. Diabetes. 2008;57(12):3247–57.
30. Lafontan M. Fat cells: afferent and efferent messages define new approaches to treat obesity. Annu Rev Pharmacol Toxicol. 2005;45:119–46.
31. Manolopoulos KN, Karpe F, Frayn KN. Gluteofemoral body fat as a determinant of metabolic health. Int J Obes (Lond). 2010;34(6):949–59.
32. Smith SR et al. Contributions of total body fat, abdominal subcutaneous adipose tissue compartments, and visceral adipose tissue to the metabolic complications of obesity. Metabolism. 2001;50(4):425–35.
33. Sbarbati A et al. Subcutaneous adipose tissue classification. Eur J Histochem. 2010;54(4):e48.
34. Krings A et al. Bone marrow fat has brown adipose tissue characteristics, which are attenuated with aging and diabetes. Bone. 2012;50(2):546–52.
35. Lecka-Czernik B. Marrow fat metabolism is linked to the systemic energy metabolism. Bone. 2012;50(2):534–9.
36. Naveiras O et al. Bone-marrow adipocytes as negative regulators of the haematopoietic microenvironment. Nature. 2009;460(7252):259–63.
37. Rajsheker S et al. Crosstalk between perivascular adipose tissue and blood vessels. Curr Opin Pharmacol. 2010;10(2):191–6.
38. Pond CM. Adipose tissue and the immune system. Prostaglandins Leukot Essent Fatty Acids. 2005;73(1):17–30.
39. Hovey RC, Aimo L. Diverse and active roles for adipocytes during mammary gland growth and function. J Mammary Gland Biol Neoplasia. 2010;15(3):279–90.
40. Wang YY et al. Adipose tissue and breast epithelial cells: a dangerous dynamic duo in breast cancer. Cancer Lett. 2012;324(2):142–51.
41. Fata JE, Werb Z, Bissell MJ. Regulation of mammary gland branching morphogenesis by the extracellular matrix and its remodeling enzymes. Breast Cancer Res. 2004;6(1):1–11.
42. Elliott BE et al. Capacity of adipose tissue to promote growth and metastasis of a murine mammary carcinoma: effect of estrogen and progesterone. Int J Cancer. 1992;51(3):416–24.
43. Manabe Y et al. Mature adipocytes, but not preadipocytes, promote the growth of breast carcinoma cells in collagen gel matrix culture through cancer-stromal cell interactions. J Pathol. 2003;201(2):221–8.

44. Dirat B et al. Unraveling the obesity and breast cancer links: a role for cancer-associated adipocytes? Endocr Dev. 2010;19:45–52.
45. Bochet L et al. Cancer-associated adipocytes promotes breast tumor radioresistance. Biochem Biophys Res Commun. 2011;411(1):102–6.
46. Park J, Euhus DM, Scherer PE. Paracrine and endocrine effects of adipose tissue on cancer development and progression. Endocr Rev. 2011;32(4):550–70.
47. Motrescu ER, Rio MC. Cancer cells, adipocytes and matrix metalloproteinase 11: a vicious tumor progression cycle. Biol Chem. 2008;389(8):1037–41.
48. Iyengar P et al. Adipocyte-secreted factors synergistically promote mammary tumorigenesis through induction of anti-apoptotic transcriptional programs and proto-oncogene stabilization. Oncogene. 2003;22(41):6408–23.
49. Khan T et al. Metabolic dysregulation and adipose tissue fibrosis: role of collagen VI. Mol Cell Biol. 2009;29(6):1575–91.
50. Motrescu ER et al. Matrix metalloproteinase-11/stromelysin-3 exhibits collagenolytic function against collagen VI under normal and malignant conditions. Oncogene. 2008;27(49): 6347–55.
51. Finley DS et al. Periprostatic adipose tissue as a modulator of prostate cancer aggressiveness. J Urol. 2009;182(4):1621–7.
52. Divoux A, Clement K. Architecture and the extracellular matrix: the still unappreciated components of the adipose tissue. Obes Rev. 2011;12(5):e494–503.
53. Subbaramaiah K et al. Obesity is associated with inflammation and elevated aromatase expression in the mouse mammary gland. Cancer Prev Res (Phila). 2011;4(3):329–46.
54. Morris PG et al. Inflammation and increased aromatase expression occur in the breast tissue of obese women with breast cancer. Cancer Prev Res (Phila). 2011;4(7):1021–9.
55. Dunlap SM et al. Dietary energy balance modulates epithelial-to-mesenchymal transition and tumor progression in murine claudin-low and basal-like mammary tumor models. Cancer Prev Res (Phila). 2012;5(7):930–42.
56. Dogan S et al. Effects of high-fat diet and/or body weight on mammary tumor leptin and apoptosis signaling pathways in MMTV-TGF-alpha mice. Breast Cancer Res. 2007;9(6):R91.
57. Zhang Y et al. White adipose tissue cells are recruited by experimental tumors and promote cancer progression in mouse models. Cancer Res. 2009;69(12):5259–66.
58. Kidd S et al. Origins of the tumor microenvironment: quantitative assessment of adipose-derived and bone marrow-derived stroma. PLoS One. 2012;7(2):e30563.
59. Chandler EM et al. Implanted adipose progenitor cells as physicochemical regulators of breast cancer. Proc Natl Acad Sci USA. 2012;109(25):9786–91.
60. Devarajan E et al. Epithelial-mesenchymal transition in breast cancer lines is mediated through PDGF-D released by tissue-resident stem cells. Int J Cancer. 2012;131(5):1023–31.
61. Zhao M et al. Multipotent adipose stromal cells and breast cancer development: think globally, act locally. Mol Carcinog. 2010;49(11):923–7.
62. Petit JY et al. Locoregional recurrence risk after lipofilling in breast cancer patients. Ann Oncol. 2012;23(3):582–8.
63. Bertolini F et al. Adipose tissue cells, lipotransfer and cancer: a challenge for scientists, oncologists and surgeons. Biochim Biophys Acta. 2012;1826(1):209–14.
64. Donnenberg VS et al. Regenerative therapy after cancer: what are the risks? Tissue Eng Part B Rev. 2010;16(6):567–75.
65. Martin-Padura I et al. The white adipose tissue used in lipotransfer procedures is a rich reservoir of CD34+ progenitors able to promote cancer progression. Cancer Res. 2012;72(1): 325–34.
66. Cannistra SA. Cancer of the ovary. N Engl J Med. 2004;351(24):2519–29.
67. Cai J et al. Fibroblasts in omentum activated by tumor cells promote ovarian cancer growth, adhesion and invasiveness. Carcinogenesis. 2012;33(1):20–9.
68. Sorensen EW et al. Omental immune aggregates and tumor metastasis within the peritoneal cavity. Immunol Res. 2009;45(2–3):185–94.

69. Smathers RL, Petersen DR. The human fatty acid-binding protein family: evolutionary divergences and functions. Hum Genomics. 2011;5(3):170–91.
70. Martinez-Outschoorn UE, Sotgia F, Lisanti MP. Power surge: supporting cells "fuel" cancer cell mitochondria. Cell Metab. 2012;15(1):4–5.
71. Gazi E et al. Direct evidence of lipid translocation between adipocytes and prostate cancer cells with imaging FTIR microspectroscopy. J Lipid Res. 2007;48(8):1846–56.
72. Cho JA et al. Exosomes from ovarian cancer cells induce adipose tissue-derived mesenchymal stem cells to acquire the physical and functional characteristics of tumor-supporting myofibroblasts. Gynecol Oncol. 2011;123(2):379–86.
73. Jeon ES et al. Cancer-derived lysophosphatidic acid stimulates differentiation of human mesenchymal stem cells to myofibroblast-like cells. Stem Cells. 2008;26(3):789–97.
74. Valet P et al. Alpha2-adrenergic receptor-mediated release of lysophosphatidic acid by adipocytes. A paracrine signal for preadipocyte growth. J Clin Invest. 1998;101(7):1431–8.
75. Jeon ES et al. Ovarian cancer-derived lysophosphatidic acid stimulates secretion of VEGF and stromal cell-derived factor-1 alpha from human mesenchymal stem cells. Exp Mol Med. 2010;42(4):280–93.
76. Protani MM, Nagle CM, Webb PM. Obesity and ovarian cancer survival: a systematic review and meta-analysis. Cancer Prev Res (Phila). 2012;5(7):901–10.
77. Schmandt RE et al. Understanding obesity and endometrial cancer risk: opportunities for prevention. Am J Obstet Gynecol. 2011;205(6):518–25.
78. Klopp AH et al. Omental adipose tissue-derived stromal cells promote vascularization and growth of endometrial tumors. Clin Cancer Res. 2012;18(3):771–82.
79. Swindle P et al. Do margins matter? The prognostic significance of positive surgical margins in radical prostatectomy specimens. J Urol. 2005;174(3):903–7.
80. Magi-Galluzzi C et al. International Society of Urological Pathology (ISUP) Consensus Conference on Handling and Staging of Radical Prostatectomy Specimens. Working group 3: extraprostatic extension, lymphovascular invasion and locally advanced disease. Mod Pathol. 2011;24(1):26–38.
81. Azevedo A et al. IL-6/IL-6R as a potential key signaling pathway in prostate cancer development. World J Clin Oncol. 2011;2(12):384–96.
82. Ribeiro RJ et al. Tumor cell-educated periprostatic adipose tissue acquires an aggressive cancer-promoting secretory profile. Cell Physiol Biochem. 2012;29(1–2):233–40.
83. Ribeiro R et al. Human periprostatic adipose tissue promotes prostate cancer aggressiveness in vitro. J Exp Clin Cancer Res. 2012;31(1):32.
84. Prantl L et al. Adipose tissue-derived stem cells promote prostate tumor growth. Prostate. 2010;70(15):1709–15.
85. Neugut AI, Chen AC, Petrylak DP. The "skinny" on obesity and prostate cancer prognosis. J Clin Oncol. 2004;22(3):395–8.
86. van Roermund JG et al. Periprostatic fat correlates with tumour aggressiveness in prostate cancer patients. BJU Int. 2011;107(11):1775–9.
87. McGuire BB, Fitzpatrick JM. BMI and the risk of renal cell carcinoma. Curr Opin Urol. 2011;21(5):356–61.
88. Stewart SB, Freedland SJ. Influence of obesity on the incidence and treatment of genitourinary malignancies. Urol Oncol. 2011;29(5):476–86.
89. Doyle SL et al. Visceral obesity, metabolic syndrome, insulin resistance and cancer. Proc Nutr Soc. 2012;71(1):181–9.
90. Notarnicola M et al. Low levels of lipogenic enzymes in peritumoral adipose tissue of colorectal cancer patients. Lipids. 2012;47(1):59–63.
91. Amemori S et al. Adipocytes and preadipocytes promote the proliferation of colon cancer cells in vitro. Am J Physiol Gastrointest Liver Physiol. 2007;292(3):G923–9.
92. Hoda MR et al. Leptin acts as a mitogenic and antiapoptotic factor for colonic cancer cells. Br J Surg. 2007;94(3):346–54.
93. Jaffe T, Schwartz B. Leptin promotes motility and invasiveness in human colon cancer cells by activating multiple signal-transduction pathways. Int J Cancer. 2008;123(11):2543–56.

94. Ratke J et al. Leptin stimulates the migration of colon carcinoma cells by multiple signaling pathways. Endocr Relat Cancer. 2010;17(1):179–89.
95. Bartucci M et al. Obesity hormone leptin induces growth and interferes with the cytotoxic effects of 5-fluorouracil in colorectal tumor stem cells. Endocr Relat Cancer. 2010;17(3): 823–33.
96. Larsson SC, Wolk A. Obesity and colon and rectal cancer risk: a meta-analysis of prospective studies. Am J Clin Nutr. 2007;86(3):556–65.
97. Moghaddam AA, Woodward M, Huxley R. Obesity and risk of colorectal cancer: a meta-analysis of 31 studies with 70,000 events. Cancer Epidemiol Biomarkers Prev. 2007;16(12): 2533–47.
98. Renehan AG et al. Body-mass index and incidence of cancer: a systematic review and meta-analysis of prospective observational studies. Lancet. 2008;371(9612):569–78.
99. Sung MK et al. Obesity-induced metabolic stresses in breast and colon cancer. Ann N Y Acad Sci. 2011;1229:61–8.
100. Park SY et al. Effects of diet-induced obesity on colitis-associated colon tumor formation in A/J mice. Int J Obes (Lond). 2012;36(2):273–80.
101. Wolin KY, Carson K, Colditz GA. Obesity and cancer. Oncologist. 2010;15(6):556–65.
102. Lysaght J et al. Pro-inflammatory and tumour proliferative properties of excess visceral adipose tissue. Cancer Lett. 2011;312(1):62–72.
103. Nomoto-Kojima N et al. Interaction between adipose tissue stromal cells and gastric cancer cells in vitro. Cell Tissue Res. 2011;344(2):287–98.
104. Muller A et al. Involvement of chemokine receptors in breast cancer metastasis. Nature. 2001;410(6824):50–6.
105. Zhao BC et al. Adipose-derived stem cells promote gastric cancer cell growth, migration and invasion through SDF-1/CXCR4 axis. Hepatogastroenterology. 2010;57(104):1382–9.
106. Calle EE et al. Overweight, obesity, and mortality from cancer in a prospectively studied cohort of U.S. adults. N Engl J Med. 2003;348(17):1625–38.
107. Jamieson NB et al. Peripancreatic fat invasion is an independent predictor of poor outcome following pancreaticoduodenectomy for pancreatic ductal adenocarcinoma. J Gastrointest Surg. 2011;15(3):512–24.
108. Balentine CJ et al. Intra-abdominal fat predicts survival in pancreatic cancer. J Gastrointest Surg. 2010;14(11):1832–7.
109. Mathur A et al. Pancreatic steatosis promotes dissemination and lethality of pancreatic cancer. J Am Coll Surg. 2009;208(5):989–94. discussion 994–6.
110. Grippo PJ et al. Concurrent PEDF deficiency and Kras mutation induce invasive pancreatic cancer and adipose-rich stroma in mice. Gut. 2012;61(10):1454–64.
111. White PB et al. Insulin, leptin, and tumoral adipocytes promote murine pancreatic cancer growth. J Gastrointest Surg. 2010;14(12):1888–93. discussion 1893–4.
112. Zyromski NJ et al. Obesity potentiates the growth and dissemination of pancreatic cancer. Surgery. 2009;146(2):258–63.
113. Cousin B et al. Adult stromal cells derived from human adipose tissue provoke pancreatic cancer cell death both in vitro and in vivo. PLoS One. 2009;4(7):e6278.
114. Bandarchi B et al. From melanocyte to metastatic malignant melanoma. Dermatol Res Pract. 2010;2010:583748.
115. Forsea AM et al. Melanoma incidence and mortality in Europe: new estimates, persistent disparities. Br J Dermatol. 2012;167(5):1124–30.
116. Villanueva J, Herlyn M. Melanoma and the tumor microenvironment. Curr Oncol Rep. 2008;10(5):439–46.
117. Gray-Schopfer V, Wellbrock C, Marais R. Melanoma biology and new targeted therapy. Nature. 2007;445(7130):851–7.
118. Smolle J et al. Pathology of tumor-stroma interaction in melanoma metastatic to the skin. Hum Pathol. 1995;26(8):856–61.
119. Wagner M et al. Inflamed tumor-associated adipose tissue is a depot for macrophages that stimulate tumor growth and angiogenesis. Angiogenesis. 2012;15(3):481–95.

120. Hollander DM et al. Demonstration of lipolytic activity from cultured human melanoma cells. J Surg Res. 1986;40(5):445–9.
121. Kushiro K, Nunez NP. Ob/ob serum promotes a mesenchymal cell phenotype in B16BL6 melanoma cells. Clin Exp Metastasis. 2011;28(8):877–86.
122. Dennis LK et al. Cutaneous melanoma and obesity in the Agricultural Health Study. Ann Epidemiol. 2008;18(3):214–21.
123. Sharma SD, Katiyar SK. Leptin deficiency-induced obesity exacerbates ultraviolet B radiation-induced cyclooxygenase-2 expression and cell survival signals in ultraviolet B-irradiated mouse skin. Toxicol Appl Pharmacol. 2010;244(3):328–35.
124. Brandon EL et al. Obesity promotes melanoma tumor growth: role of leptin. Cancer Biol Ther. 2009;8(19):1871–9.
125. Pandey V et al. Diet-induced obesity increases melanoma progression: involvement of Cav-1 and FASN. Int J Cancer. 2012;130(3):497–508.
126. Mori A et al. Severe pulmonary metastasis in obese and diabetic mice. Int J Cancer. 2006;119(12):2760–7.
127. Askmyr M, Quach J, Purton LE. Effects of the bone marrow microenvironment on hematopoietic malignancy. Bone. 2011;48(1):115–20.
128. Hino M et al. Leptin receptor and leukemia. Leuk Lymphoma. 2000;36(5–6):457–61.
129. Laharrague P et al. High expression of leptin by human bone marrow adipocytes in primary culture. FASEB J. 1998;12(9):747–52.
130. Behan JW et al. Adipocytes impair leukemia treatment in mice. Cancer Res. 2009;69(19):7867–74.
131. Gelelete CB et al. Overweight as a prognostic factor in children with acute lymphoblastic leukemia. Obesity (Silver Spring). 2011;19(9):1908–11.
132. Castillo JJ et al. Obesity but not overweight increases the incidence and mortality of leukemia in adults: a meta-analysis of prospective cohort studies. Leuk Res. 2012;36(7):868–75.
133. Medeiros BC et al. Impact of body-mass index in the outcome of adult patients with acute myeloid leukemia. Haematologica. 2012;97(9):1401–4.
134. Caers J et al. Neighboring adipocytes participate in the bone marrow microenvironment of multiple myeloma cells. Leukemia. 2007;21(7):1580–4.

Chapter 8
Trafficking of Cells from Adipose Tissue to Tumor Microenvironment

Ines Martin-Padura, Patrizia Mancuso, and Francesco Bertolini

Abstract Similar to neoplasia, the human white adipose tissue (WAT) shows in vivo a robust angiogenic switch when the growth rate exceeds a given expansion threshold. Also, antiangiogenic drugs have been found to inhibit WAT development in postnatal mice. Human WAT is very rich in CD45−CD34+ progenitors that express high levels of angiogenesis-related genes and can generate in culture endothelial cells and tubes as efficiently as mesenchymal cells. Compared to the bone marrow, WAT contains >250 times more CD45−CD34+ progenitors with endothelial differentiation potential. The coinjection of human WAT-derived CD45−CD34+ progenitors from lipotransfer procedures contributed to tumor vascularization and significantly increased tumor growth and metastases in several orthotopic models of human breast cancer in immunodeficient NSG mice. These data nicely complement the recent observation from the Kolonin laboratory that in mouse models WAT cells are mobilized and recruited by experimental tumors to promote cancer progression. Autologous lipotransfer for tissue/organ reconstruction is used in patients who had surgical removal of breast and other types of cancer. We have recently reported a study of 321 consecutive patients operated for primary breast cancer who subsequently underwent a lipotransfer procedure, compared with two matched patients with similar characteristics who did not undergo lipotransfer. In this study, the lipotransfer group exhibited a higher risk of local events compared to the controls when the analysis was limited to intraepithelial neoplasia. A second data revision after prolonged follow-up confirmed this significant difference. The dissection of the different roles of purified populations of WAT-derived progenitors and mature cells seems urgent to clarify which WAT cell populations can be used safely for tissue/organ reconstruction in cancer patients.

I. Martin-Padura • P. Mancuso • F. Bertolini, M.D., Ph.D. (✉)
Laboratory of Hematology-Oncology, European Institute of Oncology, via Ripamonti 435, 20141 Milan, Italy
e-mail: francesco.bertolini@ieo.it

M.G. Kolonin (ed.), *Adipose Tissue and Cancer*, DOI 10.1007/978-1-4614-7660-3_8,

8.1 Introduction

In the last decade, the catalytic and quantitative roles of bone marrow (BM)-derived progenitor cells in cancer growth have been intensively debated [1–12]. Donor-derived endothelial cells have been found, albeit in limited number, in patients who received allogeneic BM transplants [2]. Some investigators found a crucial and quantitatively relevant role for BM-derived vessels in the early phases of cancer growth in some preclinical models of neoplasia, some others failed to find any significant role for BM-derived endothelial progenitor cells (EPCs) in cancer vessels and in cancer development in several others preclinical models.

One study has recently described that EPCs are present in tissues other than the BM, and in particular in the adipose tissue of mice [13]. We have recently described that the human white adipose tissue (WAT) is a very rich reservoir of CD45−CD34+ EPCs. When purified human WAT-derived CD34+ cells were compared to BM-derived CD34+ cells mobilized in blood by G-CSF, purified WAT cells were found to express similar levels of stemness-related genes, significantly increased levels of angiogenesis-related genes, and increased levels of FAP-alpha, a crucial suppressor of antitumor immunity [14]. In vitro, WAT-CD34+ cells generated mature endothelial cells and endothelial tubes, and the coinjection of human WAT-CD34+ cells significantly increased tumor growth and metastases in orthotopic models of human breast cancer in NSG mice, where human WAT-CD34+ cells contributed to tumor vascularization.

8.2 Materials and Methods

8.2.1 CD34+ Cells Purification from Lipotransfer Samples

Human WAT samples were obtained from lipotransfer procedures for breast reconstruction in breast cancer patients who signed an informed consent. Most of these procedures involved WAT collection from the abdomen.

Stromal-vascular cell fractions were obtained after collagenase digestion (HBSS, Gibco, containing 2 mg/ml of collagenase type I by Sigma Aldrich and 3.5 % bovine serum albumin by Sigma Aldrich) at 37 °C with constant shaking for 60 min. The digestion was blocked with RPMI 1640 supplemented by 20 % FBS (Euroclone), and a cell pellet was obtained by centrifugation at 200×*g* for 10 min at 4 °C [15–17].

$CD34^+$ cells were purified from WAT samples by means of anti-CD34 microbeads (Miltenyi Biotec) according to the manufacturer's instructions. Final $CD34^+$ cell purity was evaluated by flow cytometry and found in each instance to be more than 95 %.

CD34+ cells obtained from WAT were investigated by flow cytometry, gene expression analysis, culture, and in vivo studies.

After informed consent was obtained, CD34+ cells were purified from WAT samples and blood apheresis products of healthy donors undergoing stem cell collection after G-CSF administration by mean of anti-CD34 microbeads (Miltenyi Biotec, Bergisch Gladbach, Germany) according to the manufacturer's instructions. Final CD34+ cell purity was evaluated by flow cytometry and found to be always >95 %.

Mouse WAT samples were obtained from the mammary and the ovary fat pads and processed as described above for human WAT.

8.2.2 *Flow Cytometry*

CD45−CD34+ progenitor cells were evaluated by six-color flow cytometry following an approach recently validated for the enumeration of circulating EPCs and perivascular progenitors [18, 19]. The nuclear staining Syto16 was used to discriminate between DNA-containing cells, platelets, and cell debris. 7AAD was used to determine the viability status of the cells. Regarding antihuman antibodies, we used anti-CD45-APC-Cy7 (clone 2D1), -CD34-APC and -PeCy7 (clone 8G12), -CD31-PeCy7 (custom product, clone L133.1), -CD13-APC (cloneWM15), -CD10-APC (clone H10a), -CD140b-PE (clone 28D4), -CD29-PE (clone MAR4), and -CD90-PE (clone 5E10) from BD (Mountain View, CA); anti-VEGFR2-PE (clone 89106) and -VEGFR3-PE (clone 54733) were from R&D Systems (Minneapolis, MN); anti-CD44-APC (clone BJ18) was from Bio-Legend (San Diego, CA); and anti-CD144-PE (TEA 1/31) was from Beckman-Coulter (Brea, CA). The nuclear staining 7-AAD and Syto 16 were from Sigma (St. Louis, MO) and Invitrogen (Carlsbad, CA), respectively. Regarding murine studies, anti-CD45-APC-Cy7 (clone 30-F11) and -CD117-PE (clone ack45) were from BD. Anti-CD13-PE (clone WM15), -Sca-1 APC (clone D7), -CD34-PC-7 (clone RAM34), and -CD150-PE (clone BD1) were from Ebioscience (San Diego, CA).

8.2.3 *RT-PCR and Expression Analysis*

In magnetically labeled CD34+ cells, RNA isolation was done using QIAmp RNA blood mini kit (Qiagen, Valencia CA) and cDNA was generated from 40 ng of RNA using the high-capacity cDNA reverse transcription kit (Applied Biosystems, Foster City, CA); quantitative PCR was made with ABI Prism 700 platform as previously described [20] using primers and probes from the TaqMan® Gene Expression Assay.

For microarray analysis, synthesis of labeled targets, array hybridization (Affymetrix GeneChip Gene ST 1.0 Human array; Affymetrix, Santa Clara, CA), staining, and scanning were performed according to Affymetrix standard protocols, starting from 500 ng of total RNA. Duplicate microarrays were hybridized with each cRNA sample. The MAS5 algorithm was used to determine the expression levels of mRNAs; the absolute analysis was performed using default parameters and scaling factor 500. Report files were extracted for each microarray chip, and performance of the labeled target was evaluated on the basis of several values (scaling factor, background, and noise values, percent present calls, average signal value, etc.). The data were deposited at GEO (http://www.ncbi.nlm.nih.gov/geo/query/acc.cgi?acc=GSE31415). Results were confirmed by quantitative RT-PCR.

8.2.4 Cell Lines and Culture

MDA-MB-436 and HCC1937 triple negative breast cancer cells were purchased from the American Type Culture Collection (ATCC, Manassas, VA, USA) and cultured as suggested by the manufacturer. Prior to injection in mice, cells (1×10^6/20 μl/mouse) were mixed with Matrigel (BD) and trypan blue solution (Sigma Aldrich, 25 % and 10 % in PBS, respectively).

Endothelial cells and capillary tubes in Matrigel were obtained from cultures of purified human WAT-CD34+ cells as previously described [9, 21]. In brief, cells were plated in complete EGM-2 medium (Lonza, Visp, Switzerland) in 12- or 24-well plates precoated with rat tail collagen I. Plates were placed in a 37 °C, 5 % CO_2 humidified incubator. The seeding density ranged from 50×10^3 to 500×10^3 cells. The presence of endothelial cells and colonies was assessed using an inverted microscope. After 3–7 days of culture, endothelial cell colonies were morphologically identified and afterward picked out using cloning rings. Fibroblast contamination was avoided by depleting them from cell suspensions with the antifibroblast microbeads kit (Miltenyi). Endothelial cell surface antigen expression was assessed by flow cytometry and immunofluorescence staining of VE-Cadherin as previously described [20, 21] (http://www.ncbi.nlm.nih.gov/geo/query/acc.cgi?acc=GSE31415).

Capillary-like structures were obtained in culture using a commercial kit (Chemicon, Temecula, CA) as previously described [21]. Briefly, matrix solution was thaw on ice, seeded on 24-well plates and incubated at 37 °C to solidify. EC were harvested, resuspended in complete media, seeded at a final concentration of 5×10^4 cells per cell onto the polymerized EC Matrix, and incubated at 37 °C in a tissue incubator. After 17 h tube formation was inspected under an inverted light microscope at 20× magnification.

Spheres were obtained in cultures as previously described [22]. In brief, cells were plated onto ultraslow attachment plates (BD-Falcon) at a density of 40,000 viable cell/ml in a serum-free mammary epithelial basal medium (MEBM, Lonza), supplemented with 5 mg/ml insulin, 0.5 mg/ml hydrocortisone, B27 (Invitrogen), 20 ng/ml EGF and bFGF (BD Biosciences), and 4 mg/ml heparin (Sigma Aldrich) and maintained in a 5 % CO_2-humidified incubator at 37 °C. Six to eight days later sphere formation was analyzed under an inverted light microscope.

8.2.5 Orthotopic Xenograft In Vivo Studies

Female NOD/SCID IL2R gamma null (NSG) mice [23, 24], 6–9 weeks old, were bred and housed under pathogen-free conditions in our animal facilities (IFOM-IEO campus, Milan, Italy). Mice were expanded from breeding pairs originally donated by Dr. Leonard Shultz. All animal experiments were carried in accordance with national and international laws and policies.

Prior to injection, tumor cells were trypsin detached, washed twice and resuspended in PBS to a final concentration of 10^6 cells/13 μl. Cell suspension was then mixed with 5 μl growth factor-reduced Matrigel (BD Biocoat) and 2 μl trypan blue solution (Sigma Aldrich) and maintained on ice until injection. In cases where tumor cells were coinjected with 2×10^5 WAT-derived cells, cell suspensions were mixed before final suspension in Matrigel. Aseptic conditions under a laminar flow hood were used throughout the surgical procedure. Mice were anesthetized with 0.2 % Avertin (Sigma Aldrich), make them laid on their back, and 20 μl cell suspension in Matrigel were injected directly in the fourth mammary fad pad through the nipple using a Hamilton syringe.

Tumor growth was monitored weekly using digital calipers and tumor volume was calculated using the formula: $L \times W^2/2 = \text{mm}^3$.

8.2.6 Bilateral Studies

NSG mice were divided into two groups, a control group in which 1×10^6 MDA-MB-436 or HCC1937 cells were injected into the right fourth mammary fat pad (through the fourth nipple) and an experimental group in which 1×10^6 MDA-MB-436 or HCC1937 cells were coinjected with 2×10^5 human CD34+ WAT cells into the right fourth mammary fat pad.

8.2.7 Monolateral Studies

1×10^6 MDA-MB-436 or HCC1937 cells were injected into the right fourth mammary fat pad and 1×10^6 MDA-MB-436 or HCC1937 cells were coinjected with 2×10^5 human CD34+ WAT cells into the left fourth mammary fat pad of the same NSG mouse.

In both set of studies, tumors were measured at least once a week using digital calipers. Tumor and lung tissues were removed at experimental end on day 84. Tumors were measured and weighed. For histological evaluation of the tumors, one part of the tumor tissue was fixed in 4 % phosphate-buffered formalin and embedded in paraffin. For detection of the pulmonary metastases, lungs were fixed in 4 % phosphate-buffered formalin and embedded in paraffin. Five micron sections of the entire lungs were made, and slides were counterstained with hematoxylin and eosin (H&E) for the detection of metastases. The Scan Scope XT device and the Aperio Digital pathology system software (Aperio, Vista, CA) were used to detect metastases.

In the second model of breast cancer metastases, 1×10^6 MDA-MB-436 cells were injected into the right fourth mammary fat pad (through the fourth nipple) of NSG mice to produce orthotopic primary tumors. When the tumor size was 200–250 mm^3, i.e., about 45 days after tumor implant, tumor resection (mastectomy) was performed. The tumor mass was gently removed and the incision closed with wound clips. Three days after mastectomy, mice were divided into an experimental group in which 2×10^5 human CD34+ WAT cells were injected into the right third mammary fat pad (through the third nipple), a group in which 2×10^5 human $CD34^-$ WAT cells were injected into the right third mammary fat pad and a control group without WAT cells injection ($n = 6$ per study group). Two months after cells injection, mice were sacrificed and right axillary lymph node and lung tissue were removed. To confirm the presence of metastases, sections were cut and stained with H&E.

For HFD studies, mice were bred and fed as previously described [25].

8.2.8 Confocal Microscopy

Images have been acquired using a Leica TCS SP5 confocal microscope and sequential Z-stacks have been performed using a 63× 1.4 NA oil immersion objective, zoom 3×, 0.3 μm Z-step. For the imaging of the red cells, the 561 laser line has been used and the autofluorescence of the cells has been collected.

8.2.9 Statistical Analysis

The Shapiro–Wilk test was used to test for normality. Considering that the very large majority of data were not normally distributed, statistical comparisons were performed using the nonparametric U-test of Mann–Whitney. All reported p-values were two sided.

8.3 Results

8.3.1 Functional Progenitor Endothelial Cells Are Stored in the White Adipose Tissue

In order to evaluate the presence of EPCs in WAT, adipose tissue was obtained from lipotransfer/lipolling procedures for breast reconstruction in breast cancer patients and processed by enzymatic digestion. By means of flow cytometry (Fig. 8.1), WAT

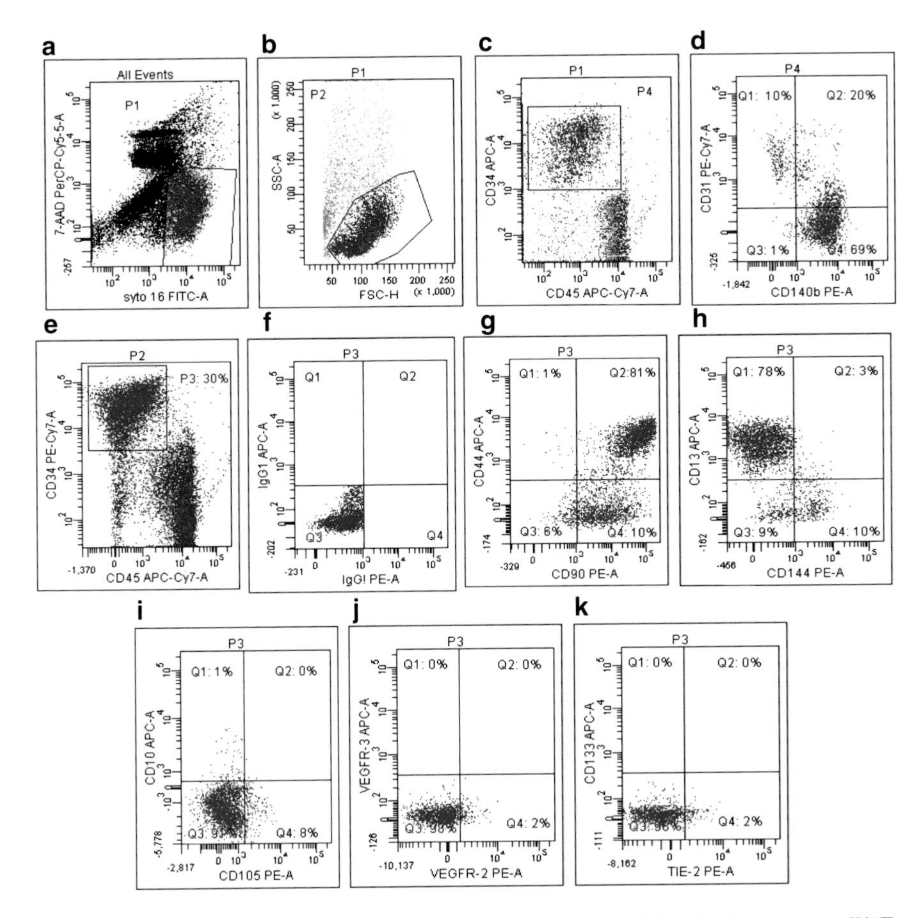

Fig. 8.1 Flow cytometry evaluation and enumeration of CD45–CD34+ cells in human WAT. Representative evaluation of CD45–CD34+ cells in human WAT tissue from lipotransfer procedures. *Panels* (**a**), (**b**), and (**e**) show the gates used to investigate DNA (Syto16)+CD34+CD45– cells. *Panel* (**f**) shows the negative controls. *Panels* (**c**), (**d**), (**g**), (**h**), (**i**), (**j**), and (**k**) show the expression of CD34, CD45, CD31, CD140b, CD31, CD44, CD90, CD13, CD144, CD10, CD105, VEGFR2, VEGFR3, CD133, and Tie2 in the DNA (Syto16)+CD34+CD45– WAT cells

was found to contain a large amount of CD45–CD34+ cells that fulfilled the most recent criteria for EPC identification [9–12] including a DNA-staining procedure to exclude contamination with platelets and/or micro- and macroparticles as previously described [18, 19]. These CD45–CD34+ cells consist of two subpopulations: $CD34^{++}CD13^{+}$ $CD140b^{+}$ $CD44^{+}$ $CD90^{++}$ cells and $CD34^{+}$ $CD31^{+}CD105^{+}$ cells. The immunomagnetic purification procedure used in the study led to a cell population, which included 79–96 % of CD45–$CD34^{++}$ $CD13^{+}$ $CD140b^{+}CD44^{+}CD90^{++}$ cells, and 2–18 % of CD45– $CD34^{+}$ $CD31^{+}CD105^{+}$ cells. CD34– cells always made up less than 5 % of the purified cell population (Fig. 8.2).

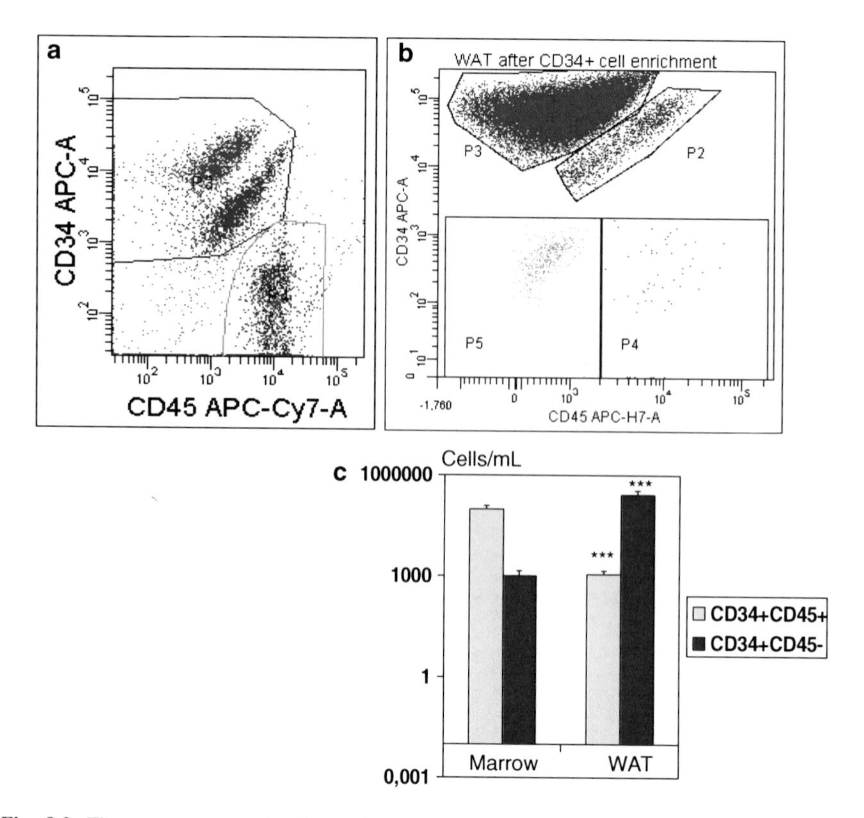

Fig. 8.2 Flow cytometry evaluation and enumeration of CD45–CD34+ cells in human WAT. Representative evaluation of CD45–CD34+ cells in human WAT tissue from lipotransfer procedures. *Panel* (**a**) shows the gate used to investigate CD34+CD45– cells. *Panel* (**b**) shows a representative evaluation of the CD34+ cell population obtained by the immunomagnetic purification procedure (79–96 % of CD45–CD34++ CD13+ CD140b+ CD44+ CD90++ cells, in *purple*; 2–18 % of CD45–CD34+ CD31+CD105+ cells, in *blue*; <5 % of CD34– cells). *Panel* (**c**) shows the quantitative enumeration of CD34+CD45+ (hematopoietic) and CD34+CD45– (endothelial) cells in human bone marrow and WAT ($n=32$). ***$p<0.0005$ vs. marrow by Mann–Whitney U-test. From Martin-Padura et al. [26], modified

Quantitative studies showed that human WAT contains about 263-fold more CD45-CD34$^+$ EPCs/ml than the BM (Fig. 8.2c, $n=32$). In particular, median human WAT CD45– CD34$^+$CD31$^+$ cells were 181,046/ml (range, 35,970–465,357), and CD45–CD34$^+$ CD31$^+$ cells were 76,946/ml (range, 13,982–191,287). Interestingly, correlations were found between the body mass index and total CD34$^+$ cells ($r=0.608$, $p<0.001$) and between WAT donor age and total CD34$^+$ cells ($r=0.387$, $p=0.035$).

In mice, EPCs are defined by surface expression of CD45$^-$ CD34$^+$ CD13$^+$ Sca1$^+$, whereas hematopoietic progenitors are defined as Lin$^-$ Sca-1$^+$CD150$^+$ cells. Enzymatic digestion of different WAT fraction, such as omentum, periovarian WAT, and breast, showed that, similar to humans, mouse WAT was also richer in EPC than was bone marrow; specifically, WAT have 179-fold more EPCs/ml than bone marrow.

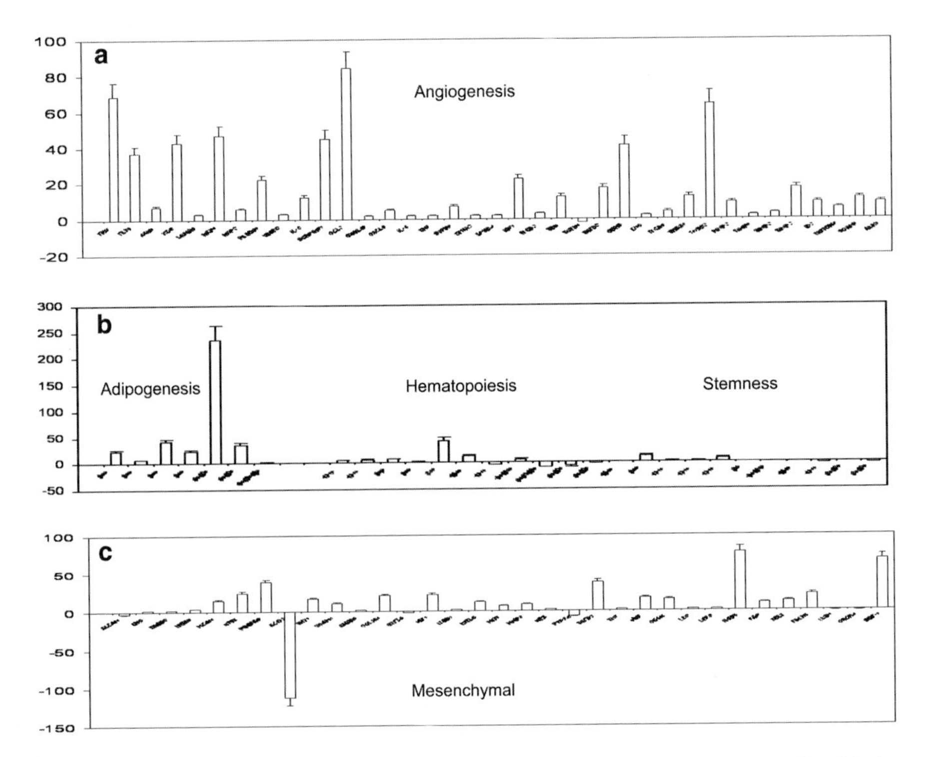

Fig. 8.3 Gene expression profile in purified CD34+ WAT cells vs. CD34+ cells mobilized in the blood by G-CSF. WAT-CD34+ cells expressed significantly higher levels of angiogenesis- (**a**) and adipogenesis- (**b**) related genes, similar levels of hematopoietic- and stemness-related genes (**b**), and higher levels of mesenchymal-related genes (**c**). From Martin-Padura et al. [26], modified

In addition to EPCs, we also investigated the presence of CD45⁺CD34⁺ hematopoietic progenitor cells in human WAT (median 6,141/ml, range <1–161,338). On average, human WAT contained 87 times less CD45⁺CD34⁺ hematopoietic progenitor cells/ml than did the BM. When seeded in methylcellulose and under the appropriate stimulation, WAT-derived human hematopoietic progenitors are able to form 1.4 ± 0.2 granulocyte macrophage colony-forming units and 3.4 ± 1.8 BFU-E/105 suggesting that cells obtained from WAT tissue are multipotential and functional cells.

8.3.2 *WAT-CD34⁺Cells Express High Levels of Angiogenesis-Related Genes and of Stemness-Related Genes*

Affymetrix gene expression profile of WAT-derived CD34⁺cells was compared with that of purified bone marrow (BM)-CD34⁺cells mobilized by G-CSF in peripheral blood of healthy donors and confirmed by qRT-PCR (Fig. 8.3). We found that series of genes associated with angiogenesis (e.g., VEGFR-1 and -2, NRP1, TEK,

VE-Cadherin, VCAM-1, and ALK1), adipogenesis (e.g., LPL, FABP4, and PPARG), endothelial, and mesenchymal differentiation (e.g., RGS5, insulin-like growth factor I, and platelet-derived growth factor receptor b) were significantly upregulated in WAT-derived $CD34^{+}$cells compared with MB–derived $CD34^{+}$cells. A panel of genes associated with stemness (e.g., SOX2, LIF, WNT3A, and Nanog) and hematopoiesis (e.g., RUNX, IL-6, and CSF-1) was expressed in WAT and bone marrow-derived $CD34^{+}$ cells at similar levels. Interestingly, WAT-derived $CD34^{+}$ cells expressed higher levels of Fibroblast activation protein-α (FAP-α) a known immune-suppressive component of the stroma [14], and of genes of the brain/adipocyte-BDNF/leptin axis, which has recently been suggested to play a relevant role in cancer progression [27].

8.3.3 *WAT-CD34⁺ Cells Are Cells with Self-Renewal Properties and Can Differentiate in Endothelial Cells In Vitro and In Vivo*

Isolated WAT-derived $CD34^{+}$ cells were grown in serum-free medium as floating spheres showing their self-renewal capacity and these spheres were found to be formed by $CD45^{-}CD13^{+}CD34^{+}$ $CD44^{+}CD90^{+}$ cells. In addition, when cultured in appropriate endothelial-differentiation media, WAT-derived $CD34^{+}$ cells generated mature endothelial cells expressing VE-cadherin that were able to organized endothelial capillary tubes when seeded in 3D Matrigel (Fig. 8.4). Moreover, orthotopic transplantation of WAT-derived cells with MDA-MB436 breast cell line gives rise to tumors (Fig. 8.5) that contained vascular vessels of human origin (labeled with antihuman CD31 species-specific antibodies in Fig. 8.6) connected with the mouse vasculature demonstrating their functionality. Confocal microscopy studies showed the presence of human $CD31^{+}CD34^{+}$ endothelial vessels in tumors of mice coinjected with breast cancer cells with a lumen where red blood cells were found.

8.3.4 *WAT-CD34⁺ Progenitors Are Increased in Mice Receiving High-Fat Diet*

Obesity is a known risk factor for breast cancer carcinoma and changes in adipose tissue structure and characteristics vary in obese patients. We hypothesized that the number of EPCs may vary when animals were fed with HFD. In fact, the number of EPCs in the mammary and ovarian WAT tissue of mice receiving an HFD ($n=12$) EPCs was significantly increased in the WAT of mice receiving an HFD but did not vary in the BM. On the contrary, hematopoietic progenitors were significantly decreased in bone marrow of mice receiving HFD (data not shown) and slightly augmented in WAT.

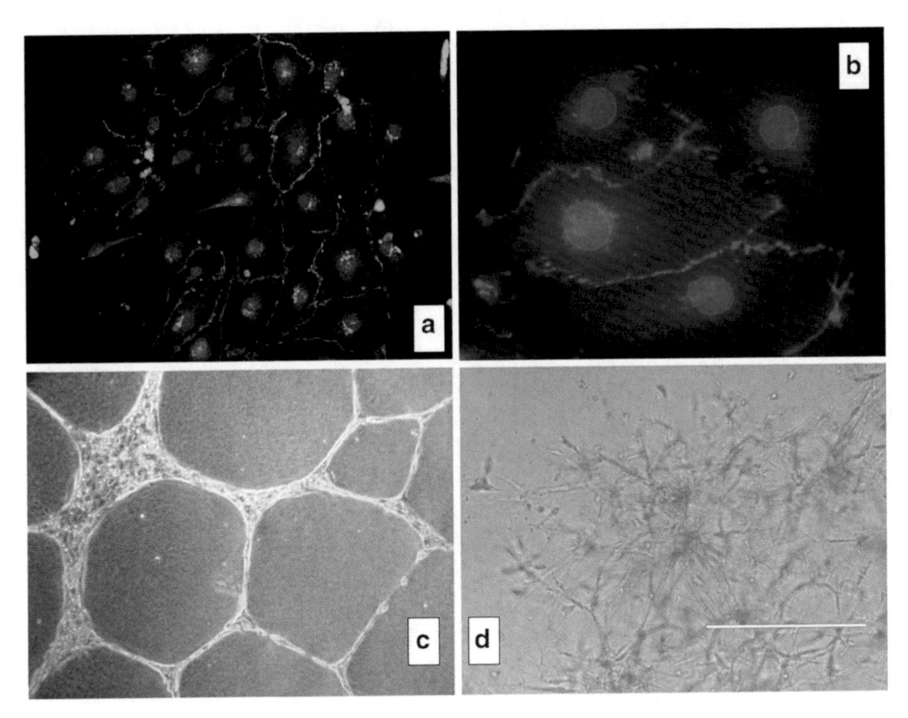

Fig. 8.4 In vitro endothelial differentiation of CD34+CD45– cells from human WAT. Clockwise from *top left*: *Panels* (**a**) and (**b**) show representative in vitro generation of endothelial cells (depicted by immunofluorescence expression of the endothelial-restricted VE-cadherin antigen) when CD34+ cells were cultured. *Panels* (**c**) and (**d**) show the in vitro generation in Matrigel of endothelial capillary tubes from CD34+ cells. From Martin-Padura et al. [26], modified

8.3.5 WAT-Derived CD34⁺ Cells Increase Tumor Growth and Metastases in Breast Cancer Models

To study the role of CD34$^+$ WAT-derived cells on tumor growth role, different orthotopic models of human breast cancer were used (Fig. 8.5). First, triple-negative human breast cancer MDA-MB-436 and HCC1937 cell lines were injected in the mammary fat pad alone or coinjected with WAT-derived human cells ($n = 124$). The coinjection of breast cancer cells and purified CD34$^+$ WAT cells significantly increased tumor growth compared to tumors obtained from transplant of tumor cells alone. Similar results were obtained in NSG mice injected with HCC1937 breast cancer cells alone or in combination with human CD34$^+$ WAT cells. WAT-derived cells alone had no tumorigenic capacity and did not grow in mice.

We conducted two separate studies to examine the in vivo involvement of CD34$^+$ WAT cells in promoting tumor growth. In bilateral studies, human CD34$^+$ WAT cells were coinjected with breast cancer cells in one of the lateral mammary fat pads (with the contralateral mammary fat pad injected with breast cancer cells alone, as control). In monolateral studies, human CD34$^+$ WAT cells were coinjected with

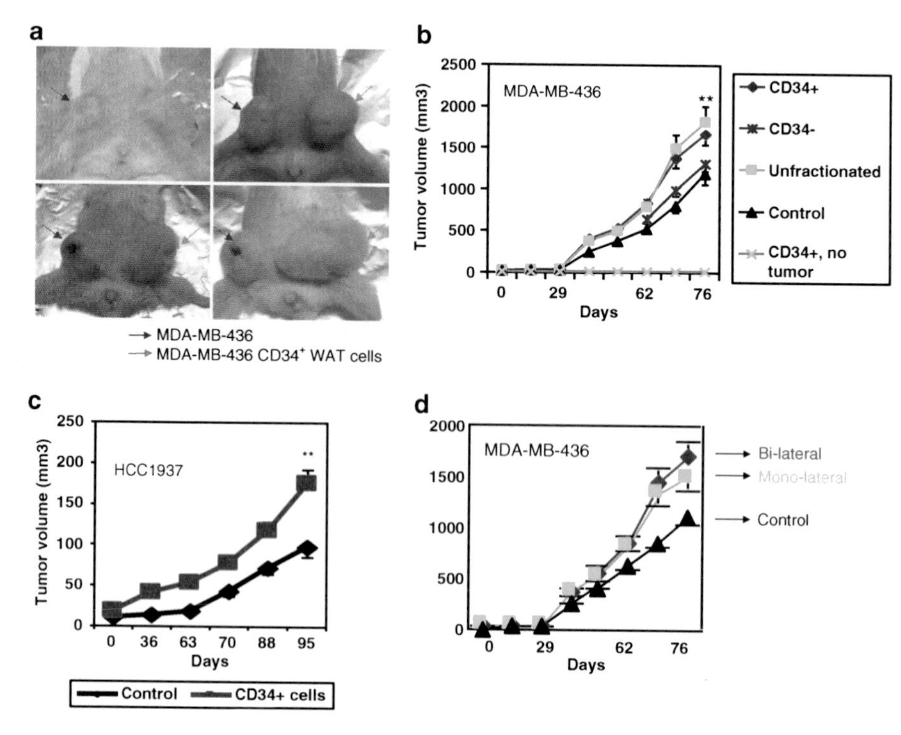

Fig. 8.5 Orthotopic models of breast cancer. *Panel* (**a**) shows representative pictures of the growth of human MDA-MB-436 breast cancer cells in the mammary fat pad of NSG mice injected with breast cancer cells alone (*red arrow*) or breast cancer cells plus CD34+ WAT cells (*blue arrow*). *Panel* (**b**) shows tumor growth in NSG mice injected with WAT-CD34+ cells alone, MDA-MB-436 cells alone, MDA-MB-436 cells plus unfractionated WAT cells, and MDA-MB-436 cells plus CD34+ or CD34− WAT cells. *Panel* (**c**) shows tumor growth in NSG mice injected with HCC1937 breast cancer cells alone and in NSG mice injected with the same number of breast cancer cells plus human CD34+ WAT cells. *Panel* (**d**) shows tumor growth in NSG mice injected with MDA-MB-436 breast cancer cells in mono- and bi-lateral studies. $^{**}p<0.005$ vs. control by Mann–Whitney U-test. From Martin-Padura et al. [26], modified

breast cancer cells in a single mammary fat pad of a series of animals (with another series of mice being injected with breast cancer cells alone in the corresponding mammary fat pads, as control). Tumor growth was slightly (albeit not significantly) higher in bilateral studies compared with monolateral studies. These data suggest that human WAT-CD34^{+}cells exert most (if not all) of their tumor-promoting activity locally and not via soluble factors that are released in circulation, which would also have promoted the growth of tumors in the opposite mammary fat pad that was not coinjected with human WAT-CD34^{+} cells.

In our model of breast cancer, where MDA-MB-436 cells were injected in the mammary fat pad, lung metastases were observed around day 70. The number of lung metastases was significantly increased in mice coinjected with breast cancer and CD34^{+}WAT cells compared with mice injected with breast cancer cells alone or mice injected with breast cancer cells and CD34^{+} WAT cells.

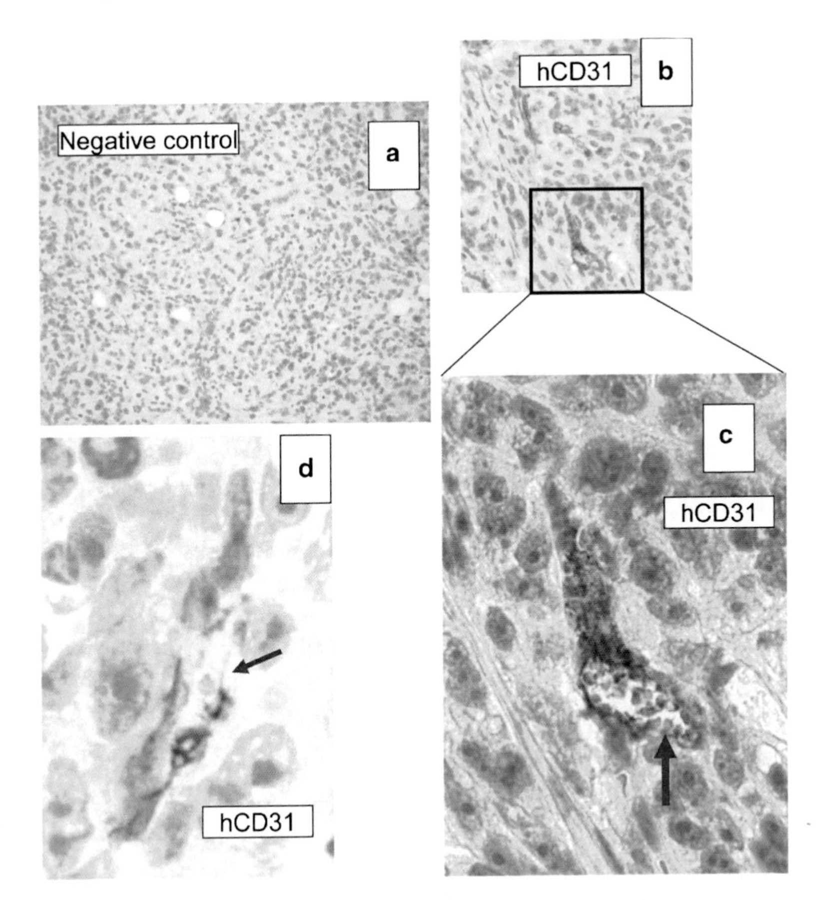

Fig. 8.6 Immunohistochemistry evaluation of human CD34+ WAT cell engraftment in breast cancer-bearing NSG mice. Clockwise, from *top left*: *panel* (**a**): negative control (human tumor in NSG mouse not injected with human WAT cells, stained with antihuman CD31). *Panels* (**b–d**) show representative evidence of the incorporation of human cells (depicted by the expression of human CD31) generated from WAT-derived CD34+ cells and lining cancer blood vessel, some of which containing red blood cells (indicated by the *red arrows*). From Martin-Padura et al. [26], modified

To mimic in a mouse model, the clinical model of lipofilling reconstruction after mastectomy we set up another model of breast cancer metastasis. MDA-MB-436 breast cancer cells were injected into the mammary fat pad of NSG mice to produce orthotopic primary tumors; when the tumor size was 200–250 mm^3, that is, about 45 days after tumor implant, the tumor was surgically removed. Mice were then divided into an experimental group in which $CD34^+$ WAT cells were injected, an experimental group in which $CD34^+$WAT cells were injected, and a control group without WAT cell injection. Two months after cell injection, mice injected with $CD34^+$WAT cells had developed an axillary metastasis and a higher number of lung metastases compared with mice injected with $CD34^-$ cells and controls ($p<0.01$).

8.4 Discussion

Our recent work about the trafficking of WAT-CD34+ progenitors offers new insight into the controversy about the quantitative and the catalytic role of EPCs in cancer growth. All previous studies investigating this topic enumerated the role of EPCs in mice carrying GFP+ (or otherwise genetically labeled) BM. This approach excluded the quantification of the role of WAT-derived EPCs that in our work were found to be in numbers significantly higher than in the BM.

Considering that our in vitro and in vivo studies were done using a cell population including both CD45–CD34+CD13+ and CD45–CD34+CD31+ cells, it is now important to better evaluate what of these two cell populations has most of the endothelial differentiation potential and of the protumorigenic and prometastatic potential. The lack of bright CD31 expression suggests that these cells are not mature endothelial cells. Moreover, as reported by Zimmerlin et al. [28], mature endothelial cells in WAT do not seem to express CD34. Along a similar line, the work of Yoder, Ingram, and Case has repeatedly indicated that the putative EPC phenotype is CD45–CD34+, and that VEGFR2 expression in immature EPCs is highly controversial in the present lack of validated reagents [29–32].

The array studies indicate that WAT-derived CD34+ cells express significantly higher levels of angiogenesis-related genes compared to BM-derived, mobilized CD34+ cells. More work with mice with GFP+ WAT is currently ongoing to further elucidate the precise role of WAT-EPCs in the tumor promotion and metastatic process.

Controversies have been reported also about the role of WAT-derived mesenchymal progenitors (MPs) in cancer growth, with some authors reporting that MPs promote tumor growth and some other reporting that MPs suppress tumor growth (reviewed in [33]). The very large majority of these studies, though, were done in mice injected with crude suspensions of MPs obtained from cell culture. To our knowledge, our study is one of the first reporting about the tumor-promoting activity of fresh human WAT-derived purified CD34+ cells. As shown by our data in mice receiving breast cancer and WAT cells only in one of the two mammary fat pads and breast cancer cells alone in the other mammary fat pad, the cancer-promoting activity of WAT-CD34+ cells is likely to be exerted through a local, rather that systemic, activity. These data nicely complement the recent observation from the Kolonin laboratory [34] that in mouse models WAT cells are mobilized and recruited by experimental tumors to promote cancer progression.

HFD was associated in mice with a significant increase in WAT-CD34+ cell numbers. HFD might interfere also with the characteristics of WAT-CD34+ cells. This, in turn, may be one of the explanations for higher incidence of breast cancer in postmenopausal obese individuals [35, 36]. So far, most of the data about the role of obesity in cancer growth focused on soluble factors, whereas our data underline the role of cellular players. In addition to EPCs, HFD increased also WAT hematopoietic progenitors. This, in turn, might increase the mobilization of hematopoietic proangiogenic cells already described by other studies [37].

A novel brain/adipocyte-BDNF/leptin axis has been recently proposed to play a potentially relevant role in cancer progression [27]. Although more studies are clearly needed to reach robust conclusions, WAT-derived CD34+ progenitors seem to express high levels of the receptors involved in this axis. Along a similar line, WAT-CD34+ cells express very high levels of FAP-alpha, a crucial suppressor of antitumor immunity [14].

Present data suggest caution about the clinical use of lipotransfer-derived WAT cells for breast reconstruction in patients with breast cancer [15, 16]. We have recently reported a study of 321 consecutive patients operated for a primary breast cancer between 1997 and 2008 who subsequently underwent a lipotransfer procedure for esthetical purpose compared with two matched patients with similar characteristics who did not undergo lipotransfer [38]. In this study, to be confirmed by prospective trials enrolling larger patients' series, the lipotransfer group resulted at higher risk of local events (four events) compared to the control group (no event) when the analysis was limited to intraepithelial neoplasia. A second data revision after prolonged follow-up confirmed this significant difference. The dissection of the different roles of purified populations of WAT-derived progenitors and mature cells seems urgent to clarify which WAT cell populations can be used safely for tissue/organ reconstruction in cancer patients.

Acknowledgment Supported in part by AIRC (Associazione Italiana per la Ricerca sul Cancro), Fondazione Umberto Veronesi, ISS (Istituto Superiore di Sanità), and Ministero della Salute. F. Bertolini is a scholar of the US National Blood Foundation.

References

1. Stoll BR, Migliorini C, Kadambi A, Munn LL, Jain RK. A mathematical model of the contribution of endothelial progenitor cells to angiogenesis in tumors: implications for antiangiogenic therapy. Blood. 2003;102:2555–61.
2. Peters BA, Diaz LA, Polyak K, et al. Contribution of bone marrow-derived endothelial cells to human tumor vasculature. Nat Med. 2005;11:261–2.
3. Ribatti D, Nico B, Crivellato E, Vacca A. Endothelial progenitor cells in health and disease. Histol Histopathol. 2005;20:1351–8.
4. Bertolini F, Shaked Y, Mancuso P, Kerbel RS. The multifaceted circulating endothelial cell in cancer: towards marker and target identification. Nat Rev Cancer. 2006;6:835–45.
5. Kaplan RN, Rafii S, Lyden D. Preparing the "soil". The premetastatic niche. Cancer Res. 2006;66:11089–93.
6. De Palma M, Naldini L. Role of hematopoietic cells and endothelial progenitors in tumor angiogenesis. Biochim Biophys Acta. 2006;1766:159–66.
7. Seandel M, Butler J, Lyden D, Rafii S. A catalytic role for proangiogenic marrow-derived cells in tumor neovascularisation. Cancer Cell. 2008;13:181–3.
8. Purhonen S, Palm J, Rossi D, et al. Bone marrow-derived circulating endothelial precursors do not contribute to vascular endothelium and are not needed for tumor growth. Proc Natl Acad Sci USA. 2008;105:6620–5.
9. Hirschi KK, Ingram DA, Yoder MC. Assessing identity, phenotype, and fate of endothelial progenitor cells. Arterioscler Thromb Vasc Biol. 2008;28:1584–95.

10. Yoder MC, Ingram DA. The definition of EPCs and other bone marrow cells contributing to neoangiogenesis and tumor growth: is there common ground for understanding the roles of numerous marrow-derived cells in the neoangiogenic process? Biochim Biophys Acta. 2009;1796:50–4.
11. Shaked Y, Voest EE. Bone marrow derived cells in tumor angiogenesis and growth: are they the good, the bad or the evil? Biochim Biophys Acta. 2009;1796:1–4.
12. Bertolini F, Mancuso P, Braidotti P, Shaked Y, Kerbel RS. The multiple personality disorder phenotype(s) of circulating endothelial cells in cancer. Biochim Biophys Acta. 2009;1796: 27–32.
13. Grenier G, Scimè A, Le Grand F, et al. Resident endothelial precursors in muscle, adipose, and dermis contribute to postnatal vasculogenesis. Stem Cells. 2007;25:3101–10.
14. Kraman M, Bambrough PJ, Arnold JN, et al. Suppression of antitumor immunity by stromal cells expressing fibroblast activation protein-alpha. Science. 2010;330:827–30.
15. Lohsiriwat V, Curigliano G, Rietjens M, Goldhirsch A, Petit YV. Autologous fat transplantation in patients with breast cancer: "silencing" or "fueling" cancer recurrence? Breast. 2011;20: 351–7.
16. Petit JY, Clough K, Sarfati I, Lohsiriwat V, de Lorenzi F, Rietjens M. Lipotransfer in breast cancer patients: from surgical technique to oncologic point of view. Plast Reconstr Surg. 2010;126:262–3.
17. Sengenès C, Lolmède K, Zakaroff-Girard A, Busse R, Bouloumié A. Preadipocytes in the human subcutaneous adipose tissue display distinct features from the adult mesenchymal and hematopoietic stem cells. J Cell Physiol. 2005;205:114–22.
18. Mancuso P, Antoniotti P, Quarna J, et al. Validation of a standardized method for enumerating circulating endothelial cells and progenitors: flow cytometry and molecular and ultrastructural analyses. Clin Cancer Res. 2009;15:267–73.
19. Mancuso P, Martin-Padura I, Calleri A, et al. Circulating perivascular progenitors, a target of PDGFR inhibition. Int J Cancer. 2011;129:1344–50.
20. Rabascio C, Muratori E, Mancuso P, et al. Assessing tumor angiogenesis: increased circulating VE-cadherin RNA in patients with cancer indicates viability of circulating endothelial cells. Cancer Res. 2004;15:4373–7.
21. Corada M, Liao F, Lindgren M, et al. Monoclonal antibodies directed to different regions of vascular endothelial cadherin extracellular domain affect adhesion and clustering of the protein and modulate endothelial permeability. Blood. 2001;97:1679–84.
22. Cicalese A, Bonizzi G, Pasi CE, et al. The tumor suppressor p53 regulates polarity of self-renewing divisions in mammary stem cells. Cell. 2009;138:1083–95.
23. Shultz LD, Ishikawa F, Greiner DL. Humanized mice in translational biomedical research. Nat Rev Immunol. 2007;7:118–30.
24. Agliano A, Martin-Padura I, Marighetti P, et al. Human acute leukemia cells injected in NOD/LtSz-scid/IL-2Rgamma null mice generate a faster and more efficient disease compared to other NOD/scid-related strains. Int J Cancer. 2008;123:2222–7.
25. Napoli C, Martin-Padura I, De Nigris F, et al. Deletion of the p66Shc longevity gene reduces systemic and tissue oxidative stress, vascular cell apoptosis, and early atherogenesis in mice fed a high-fat diet. Proc Natl Acad Sci USA. 2003;100:2112–6.
26. Martin-Padura I, Gregato G, Marighetti P, et al. The white adipose tissue used in lipotransfer procedures is a rich reservoir of CD34+ progenitors able to promote cancer progression. Cancer Res. 2012;72:325–34.
27. Cao L, Liu X, Lin EJ, et al. Environmental and genetic activation of a brain-adipocyte BDNF/leptin axis causes cancer remission and inhibition. Cell. 2010;142:52–64.
28. Zimmerlin L, Donnenberg VS, Pfeifer ME, et al. Stromal vascular progenitors in adult human adipose tissue. Cytometry A. 2010;77:22–30.
29. Ingram DA, Mead LE, Tanaka H, et al. Identification of a novel hierarchy of endothelial progenitor cells using human peripheral and umbilical cord blood. Blood. 2004;104:2752–60.
30. Ingram DA, Mead LE, Moore DB, Woodard W, Fenoglio A, Yoder MC. Vessel wall-derived endothelial cells rapidly proliferate because they contain a complete hierarchy of endothelial progenitor cells. Blood. 2005;105:2783–6.

31. Case J, Mead LE, Bessler WK, et al. Human CD34+AC133+VEGFR-2+ cells are not endothelial progenitor cells but distinct, primitive hematopoietic progenitors. Exp Hematol. 2007;35:1109–18.
32. Estes ML, Mund JA, Ingram DA, Case J. Identification of endothelial cells and progenitor cell subsets in human peripheral blood. Curr Protoc Cytom. 2010;9:1–11.
33. Klopp AH, Gupta A, Spaeth E, Andreeff M, Marini 3rd F. Concise review: dissecting a discrepancy in the literature: do mesenchymal stem cells support or suppress tumor growth? Stem Cells. 2011;29:11–9.
34. Zhang Y, Daquinaq A, Traktuev DO, et al. White adipose tissue cells are recruited by experimental tumors and promote cancer progression in mouse models. Cancer Res. 2009;69: 5259–66.
35. Harris HR, Willet WC, Terry KL, Michels KB. Body fat distribution and risk of premenopausal breast cancer in the Nurses' Health Study II. J Natl Cancer Inst. 2011;103:273–8.
36. Sinicrope FA, Dannenberg AJ. Obesity and breast cancer prognosis: weight of the evidence. J Clin Oncol. 2011;29:4–7.
37. Bellows CF, Zhang Y, Simmons PJ, Khalsa AS, Kolonin MG. Influence of BMI on level of circulating progenitor cells. Obesity (Silver Spring). 2011;19:1722–6.
38. Petit JY, Botteri E, Lohsiriwat V, et al. Locoregional recurrence risk after lipotransfer in breast cancer patients. Ann Oncol. 2012;23:582–8.

Chapter 9
The Impact of Obesity Intervention on Cancer: Clinical Perspectives

Ted D. Adams, Jessica L.J. Greenwood, and Steven C. Hunt

Abstract Previous chapters have clearly demonstrated the underlying association between obesity and risk of specific cancer types, with particular attention focused on the role of adipose tissue in cancer development. In addition to cancer incidence risk attributed to obesity, increasing cancer recurrence risk may also be associated with obesity (Ligibel, Oncology (Williston Park) 25(11):994–1000, 2011; Protani et al., Breast Cancer Res Treat 123(3):627–635, 2010; Patlak & Nass, The role of obesity in cancer survival and recurrence: workshop summary, The National Academies Press, Washington, DC, 2012; Hewitt et al., From cancer patient to cancer survivor: lost in the transition. Institute of Medicine and National Research Council, Washington, DC, 2005; Demark-Wahnefried et al., Cancer Epidemiol Biomarkers Prev 21(8): 1244–1259, 2012). Given the positive link between obesity and increased cancer risk, it should follow that cancer risk reduction could be achieved through intervention efforts to prevent patients from becoming overweight or obese or through weight loss in patients already identified as overweight or obese. Unfortunately, evidence supporting a favorable link between voluntary weight loss and subsequent reduced cancer risk is limited and not as conclusive as research findings that clearly demonstrate greater cancer risk with obesity. This chapter highlights

T.D. Adams, Ph.D., M.P.H. (✉)
Cardiovascular Genetics Division, University of Utah School or Medicine (TDA, SCH),
Salt Lake City, UT, USA
e-mail: ted.adams@utah.edu

LiVeWell Center Salt Lake City, Intermountain Healthcare, Salt Lake City, UT, USA

J.L.J. Greenwood, M.D., M.P.H.
Department of Family and Preventive Medicine, University of Utah School
of Medicine, Salt Lake City, UT, USA

S.C. Hunt, Ph.D.
Cardiovascular Genetics Division, University of Utah School or Medicine,
Salt Lake City, UT, USA

M.G. Kolonin (ed.), *Adipose Tissue and Cancer*, DOI 10.1007/978-1-4614-7660-3_9,

the proposed clinical approaches for treating adult obesity, including bariatric surgery, and explores the degree to which these obesity interventions have been shown to favorably impact cancer incidence and cancer recurrence.

9.1 Obesity and Clinical Intervention Strategy

In this chapter, adult overweight is defined as a measured body mass index (BMI) equal to or greater than 25 kg/m^2 and adult obesity is defined as a measured BMI equal to or greater than 30 kg/m^2, with obesity subcategories: Class 1 obesity, 30–34.9 kg/m^2; Class 2 obesity, 35–39.9 kg/m^2; and Class 3 obesity (extreme obesity) ≥40 kg/m^2 [1, 2]. Childhood and adolescent overweight is defined as age- and gender-specific BMI at ≥85th–94th percentile and obesity for this population is defined as age- and gender-specific BMI at ≥95th percentile.

Recommended clinical treatment for overweight and obesity (i.e., all classes) includes lifestyle therapy (diet, physical activity, and behavioral modification). For individuals whose BMI is ≥27 kg/m^2 and with two or more obesity-related risk factors or with a BMI≥30 kg/m^2, pharmacotherapy is also recommended as a possible adjunct to lifestyle therapy [3]. For patients whose weight loss attempts have failed and whose BMI is ≥35 kg/m^2 with at least two obesity-related risk factors such as hypertension and diabetes or with a BMI≥40 kg/m^2, the additional option of bariatric (weight loss) surgery has been approved for clinical intervention [3]. Recently, the US Food and Drug Administration authorized the LAP-BAND Adjustable Gastric Banding System (Allergan©) as a clinical treatment option for patients whose BMI is ≥30 kg/m^2 with at least one preexisting obesity-related risk factor.

As indicated, lifestyle therapy serves as a foundational aspect of overweight and obesity intervention and has also been identified as a key strategy to prevent overweight and obesity. The evidence-based application of lifestyle therapy in the prevention of cancer has been carefully reviewed and specific lifestyle-related recommendations have been provided in two important documents: The *American Cancer Society Guidelines on Nutrition and Physical Activity for Cancer Prevention* (American Cancer Society) [4] and *Food, Nutrition, Physical Activity and the Prevention of Cancer: A Global Perspective* (World Cancer Fund and American Institute for Cancer Research) [5]. Both documents have highlighted the importance of adhering to a healthy diet and of participating in regular physical activity for the prevention and treatment of overweight and obesity. These two documents also conform to other prevention-based guidelines proposed by the following organizations: the European Code Against Cancer for cancer prevention [6]; the American Heart Association for coronary heart disease prevention [7]; the American Diabetes Association for diabetes prevention [8]; and guidelines aimed at promoting overall good health (the 2010 *Dietary Guidelines for Americans* [9] and the 2008 *Physical Activity Guidelines for Americans* [4, 10]).

With specific reference to cancer prevention, the World Cancer Research Fund and the American Institute for Cancer Research has recommended that individuals should "be as lean as possible within the normal range of body weight." They have also recommended individuals who have gained weight, but remain within the normal weight range, work toward returning to their original weight, or individuals lose enough weight to approach the normal weight range if they are above the normal weight range [5]. Similarly, with regard to body weight and cancer prevention, the American Cancer Society's (ACS) guidelines recommend individuals: achieve and maintain weight throughout life; be as lean as possible throughout life without being underweight; and avoid excess weight gain at *all ages* [4]. For those currently overweight or obese, the ACS indicates that losing even a small amount of weight has health benefits. Although not specifically identified, the ACS's reference to "health benefits" most likely includes cancer risk reduction. Therefore, these two important national and international documents, which have specifically focused on cancer prevention guidelines, imply that through voluntary weight loss cancer risk can be reduced.

To adequately evaluate the impact of voluntary weight loss and subsequent cancer risk, reason would suggest that a significant length of follow-up time (i.e., perhaps years) coupled with a meaningful degree of sustained weight loss (i.e., perhaps at least 10 % of initial weight) would be required. With reference to weight loss studies whose primary clinical intervention strategy is lifestyle-based (i.e., diet, physical activity and behavioral modification), long-term weight loss maintenance in sizable population studies is difficult to achieve [11–13]. Large population, multicenter randomized clinical trials that have demonstrated successful weight loss through behavioral and lifestyle interventions have included intensive counseling and behavioral intervention as well as inclusion of medication. Two examples are the Diabetes Prevention Program (DDP) study [14] and the Action for Health in Diabetes (Look AHEAD) study [15, 16]. Inclusion criteria for DDP study participants (N=3,234) included increased fasting glucose (prediabetic range; 5.3–6.9 mmol/L), impaired glucose tolerance (2-h postload glucose 7.8–11.0 mmol/L), and a body mass index of 24 kg/m^2 (≥22 kg/m^2 for Asian Americans). Participants were randomly assigned to one of three groups: intensive lifestyle intervention (diet and physical activity) aimed to achieve and maintain at least a 7 % weight loss; metformin 850 mg twice per day; or placebo. The metformin and placebo groups were masked with regard to treatment. Primary study outcomes included development of diabetes with diabetes risk factors included as secondary outcomes. When study results showed that the intensive lifestyle group and the metfromin group had a 58 % and 31 % reduction in diabetes incidence, respectively, compared to the placebo group, the masked treatment was discontinued [17]. The reported weight loss patterns for the intensive lifestyle group over a 10-year period since randomization showed an initial mean weight loss of 7 kg (approximately 7.5 % loss from their initial weight) by 1 year with a gradual regain of 5 kg over the next approximately 4 years, followed by a 5-year maintenance of about 2 kg less than their mean weight at the start of the study [17].

Look AHEAD represents the first randomized control trial to explore whether or not weight loss, in combination with physical activity, results in a reduction of cardiovascular morbidity and mortality [16, 18]. Inclusion criteria for participants (N=5,145) included overweight status and a diagnosis of type 2 diabetes, and participants were randomized to either a usual care group or to an intensive lifestyle group whose goal was to lose at least 7 % of their initial weight. The group receiving the intense lifestyle intervention was also provided the option of taking a weight loss medication, Orlistat, after the first 6 months of lifestyle intervention if they had not lost at least 5 % of their initial weight. At the end of year one, the intense lifestyle group had lost on average 8.6 % of their initial weight compared to the usual care group who lost 0.7 % of their weight. At 4 years, the average weight loss of the intense lifestyle group was 6.2 % compared to −0.9 % for the controls ($p<0.001$). Further, diabetes status improved significantly among the intense treatment group (hemoglobin A1c levels (−0.36 % vs. −0.09 %; $p<0.001$) [19].

Two important points relevant to the topic of this chapter emerge from these two population studies. First, in order to achieve a meaningful and sustained weight loss using lifestyle-based intervention, and in some cases weight loss medication, an intense approach appears to be required. Given the results of these two studies, the expected degree of weight loss with intense lifestyle intervention appears to be approximately 7–9 % of initial body weight with tendency toward regain after the first year. For the DPP and Look AHEAD trials, the weight loss at 1 year was 7.5 % (DPP) and 8.6 % (Look AHEAD) from baseline weight and at 4 years weight loss was 2.1 % (DPP) and 6.2 % (Look AHEAD) from baseline [17, 19]. Second, modest weight loss has been shown to result in improvements in insulin sensitivity and sex- and metabolic-related hormones and inflammatory markers, all of which have been proposed to be associated with mechanisms linking obesity and cancer risk [4, 20, 21]. These findings raise the question of whether or not a weight loss of <10 % of initial weight is sufficient to effect physiological changes leading to reduced cancer risk. Clearly, additional research focused on the magnitude and duration of weight loss necessary to effect cancer risk reduction is needed.

9.2 Clinical Intervention for Obesity and Cancer Risk

Despite the growing body of scientific literature obtained from prospective and observational studies suggesting that clinical intervention for the treatment of obesity can have a favorable impact on lowering cancer incidence and cancer recurrence, the evidence has been limited and incomplete [4, 11, 13, 20, 22, 23]. As previously alluded, the primary impediment to ascertaining the impact weight loss has upon cancer risk has been the absence of sustained weight loss as well as the limited degree of weight lost in population-based studies [11, 13, 22]. That is, although traditional weight loss therapies, such as diet intervention, have shown short-term success with regard to weight loss, estimates suggest that perhaps as few as 5–10 % of participants achieve long-term reduction [11, 12]. Additional barriers to understanding the association between weight loss and cancer risk are the absence within the study methods to define weight loss intentionality (i.e., did study participants lose their weight voluntarily or not) and that many studies have not been initiated with weight loss as the primary intervention [11]. With these research limitations in mind, the remainder of this chapter focuses on research findings related to clinical interventions for overweight and obesity, including subsequent cancer risk outcomes when cancer-related data have been identified. While some studies may be cited where intentionality of weight loss is not known, no studies will be reviewed if weight loss is clearly defined as nonvoluntary. In addition, the potential benefits of preventing overweight and obesity (i.e., through diet and physical activity) as it relates to cancer incidence will not be discussed within this chapter, as these central topics have been clearly reviewed in preceding chapters.

Research that has focused on weight loss over an individual's lifetime has suggested a possible link between reduced cancer risk following weight loss [23–36]. In addition, during the past few years, reviews that have addressed the association between weight loss and cancer and have not exclusively focused on weight loss through bariatric surgery have been published. A review published in 2008 by Wolin and Colditz explored evidence relating both weight loss and weight gain to cancer incidence [13]. Their review concentrated on colon, breast, prostate, esophageal, pancreatic, endometrial, kidney, and renal cell cancers. Their review highlighted several studies that favored an association between weight gain and some cancers, but the degree of evidence linking weight loss to a reduction in cancer risk was limited [13]. The review identified studies that have shown a reduced risk in postmenopausal breast cancer following weight loss. They cited that limited evidence exist linking reduced prostate cancer risk and weight loss. In addition, because weight loss has been shown to lower the risk for gastroesophageal reflux, a potential partner in the mechanistic development of adenocarcinoma of the esophagus, the authors speculated that weight loss may reduce this cancer [13]. Although the review clearly identified there are limited data on weight loss, likely the result of "small numbers of individuals able to achieve sustained weight loss," the authors concluded their review with the following statement:

> "If individuals achieve and maintain weight loss, we could prevent substantial cancer burden. This is most evident for postmenopausal breast cancer. The time frame for the benefits of reduced cancer risk after successful weight loss remains unclear for most cancers." [13]

A second review by Byers and Sedjo, published in 2011, reviewed evidence related to intentional weight loss and subsequent cancer risk [20]. Aside from bariatric surgery studies, authors identified three cohort studies and three dietary randomized trials that correlated intentional weight loss with cancer risk reduction. Of these investigations, all but one cohort study focused on breast cancer as the primary outcome measure with the remaining cohort study exploring all cancer sites [20]. Reported cancer risk reduction ranged from 4 % to as high as 64 %. Studies demonstrating meaningful changes in probable cancer-related hormonal biomarkers and proinflammatory agents following intentional weight loss studies, with probable mechanistic pathways, were also identified in this review. Insight into the opportunity for future research activities were identified in this review:

> "The extent to which excess cancer risk from obesity might be reversed with intentional weight loss is unknown. There are many important questions that are not yet answered, such as how much weight needs to be lost to reduce obesity-related cancer risk, what methods of weight loss are most effective for reducing risk, just how much of the obesity-caused cancer risk can be reduced with weight loss, what latencies exist in cancer risk reduction with intentional weight loss, and how much will benefits of weight loss be offset by any subsequent regain?"

Despite the multiple questions and uncertainty identified in the above statement, the authors concluded that the data included within their review supported the "hypothesis" that intentional weight loss can "substantially reduce cancer risk," and that even a modest degree of weight loss may favorably reduce cancer risk [20].

The most recent review by Birks et al., published in 2012, systematically examined the impact of weight loss on both cancer incidence and cancer mortality [11]. Searching peer-reviewed articles published between 1978 and April 2011, and using clear definitions for weight loss and cancer as well as data extraction and quality assessment criteria, the authors identified 34 articles that complied with the inclusion criteria. These studies (all published between 1995 and 2011) were divided into four separate categories for more careful review (1) bariatric surgery weight loss and cancer ($n=3$) (these will be discussed in a separate section of this chapter); (2) intentional nonsurgical weight loss and cancer ($n=3$); (3) any weight loss (intentionality not determined) and postmenopausal breast cancer ($n=10$); and (4) any weight loss (intentionality not determined) and any cancer with the exception of postmenopausal breast cancer ($n=6$ for all cancers explored and $n=12$ for other specified cancers) [11]. From this very thorough review, Birks et al. noted that data from all but 1 of the 34 articles were from observational studies and that unfortunately, most of the investigations relied upon self-reported weights and failed to delineate the intentionality of the weight loss reported. Sixteen of the 34 papers identified a significant favorable association with weight loss and subsequent reduction in cancer incidence or cancer mortality, with the remainder of the studies reporting null findings. Baseline weight of the population under investigation was a "significant confounder." Their recommendation for future study methodology was to clearly define the intentionality of weight loss as well to stratify the baseline obesity status of participants. Therefore, it was not surprising that Birks et al. found that for studies whose methodology focused on intentional weight loss, the weight loss link with reduced cancer risk was "more consistently seen" (five of six studies) and that "the risk reduction was greatest for obesity-related cancers and in women" [11]. The authors concluded:

> "…intentional weight loss does result in a decreased incidence of cancer, particularly female obesity-related cancers. However, there is a need for further evaluation of sustained intentional weight loss in the obese and with less reliance on self-reported weight data and more focus on male populations."

With reference to the impact of clinical intervention for overweight and obesity through the use of weight loss-specific pharmacological agents and their impact upon cancer risk, the opinion of the authors of this chapter is that there is very little conducted research, if any, addressing this possible association. In support of this belief, we note that there are a limited number of weight loss medications now approved by the US Food and Drug Administration, and most of these are approved for only short-term use (i.e., 12 weeks), although physicians prescribe these medications off-label. In addition, effectiveness of long-term weight loss through pharmacological therapy has been marginal, with an initial reported weight loss of approximately 10 pounds more than might have been achieved without drug therapy, with most research demonstrating that after patients have discontinued taking weight loss medications, they regain the weight they initially lost. Clearly, additional studies are needed to assess the long-term effects that weight loss medication use has upon health, including potential links with cancer risk. Although this chapter does not explore the possible association that certain medications used as therapy for chronic illnesses might have with subsequent cancer risk (i.e., metformin uses for the treatment of diabetes), we note that research related to this potential link is ongoing. For example, a recent literature search and meta-analysis (2010) of epidemiologic studies that had been conducted for the purpose of assessing the effects of metformin therapy on cancer incidence as well as cancer mortality in diabetic patients, demonstrated a 31 % reduction in cancer incidence with an "overall summary relative risk (0.69; 95 % confidence interval, 0.61–0.79)" for subjects whose therapy was metformin compared to other antidiabetic medications [37]. The authors further concluded that their analyses noted "promising trends" for overall cancer mortality and "on specific cancer sites, particularly pancreatic cancer and hepatocellular carcinoma (HCC) and, to a lesser extent, colon and breast cancers" [37].

9.3 Weight Loss Through Bariatric Surgery

As previously highlighted, bariatric surgery remains a clinical treatment option for patients with: failed nonsurgical attempts at weight loss; a BMI $\geq$35 kg/m^2 but <40 kg/m^2 and at least two obesity-related risk factors; or a BMI $\geq$40 kg/m^2. The adjustable Gastric Banding System (Allergan©) is also a clinical treatment option for patients whose BMI is $\geq$30 kg/m^2 and have at least one preexisting obesity-related risk factor. Given the fact that the prevalence of extreme obesity in the USA is increasing at a rate greater than moderate obesity [38, 39], it is not surprising that surgical options for treatment of the severely obese population have increased in popularity over the last few decades, with an estimated 344,000 cases performed globally in 2008 [40, 41]. As previously noted, lifestyle therapy for weight loss intervention is generally insufficient for extremely obese patients and effective long-term weight loss using pharmacological therapy has been limited, leaving bariatric surgery as the only medical intervention providing substantial, long-term weight loss for most severely obese patients. Perhaps the most well-known surgical weight loss study is the prospective Swedish Obesity Subjects (SOS) study in which post-bariatric patients have demonstrated significant and sustained weight loss (i.e., greater than 10 years) when compared to matched severely obese control participants [42]. To date, three randomized clinical trials have been published comparing diabetic patients with bariatric surgical procedures or intensive medical therapy [43–45]. Dixon et al. performed a randomized, unblinded, controlled trial in which recently diagnosed diabetic patients (<2 years) whose BMI was 30–40 kg/m^2 were randomized to undergo laparoscopic adjustable gastric banding surgery or participate in a nonsurgical medical intervention program for weight loss [45]. After 2-year intervention, the bariatric surgery group had lost 20.7 % of their initial body weight compared to 1.7 % in the medical intervention group. Dixon et al. also reported that after 2 years, type 2 diabetes remission for the gastric banding group was 73 % compared with 13 % in the conventional therapy group [45]. Mingrone et al. randomized patients to bariatric surgery (gastric bypass or biliopancreatic diversion) or conventional medical therapy group [43]. At 2-year follow-up, weight losses from baseline weight were 33.3 %, 33.8 %, and 4.7 % for the gastric bypass, biliopancreatic diversion, and intensive lifestyle therapy program, respectively. Using the definition for remission of diabetes proposed by Buse et al. [46], Mingrone et al. found diabetes remission rates at 2 years were 75 % for gastric bypass, 95 % for biliopancreatic diversion, and no remission for the conventional medical therapy group [43]. Finally, Schaeur et al., after 1 year reported weight loss from baseline weight of 27.5 % for gastric bypass patients, 24.7 % for sleeve patients, and 5.2 % for the intensive lifestyle-based program. Schauer et al. also reported 42 %, 37 % and 12 % of gastric bypass, sleeve gastrectomy, and medical therapy groups, respectively, achieved the primary endpoint of a glycated hemoglobin level of 6 % or less [44]. Because post-bariatric surgical patients generally experience significant and sustained weight loss [2, 47], they represent a unique population to study the relationship between voluntary weight loss and cancer risk.

Bariatric surgery can be performed using an "open" procedure approach where the operation is conducted through an abdominal incision, or by means of a laparoscope. Laparoscopic-performed bariatric surgery is by far the most commonly performed today as patients experience fewer postsurgical complications such as surgery-related hernia and earlier hospital discharges. However, for some patients, the laparoscopic approach is not indicated. Potential contraindications include those patients with an extremely high BMI, with previous stomach surgical procedures, or in some cases, with complex medical histories [48]. There are essentially four types of bariatric surgery, although multiple variations of these primary procedures are performed and have been reported in the clinical literature. These operations include adjustable gastric band (AGB), Roux-en-Y gastric bypass (RYGB), vertical sleeve gastrectomy (VSG), and the biliopancreatic diversion with a duodenal switch (BPD-DS, often referred to as the "duodenal switch") (see Fig. 9.1). These operations are often further classified as purely restrictive such as the AGB, or malabsorptive or both such as the RYGB, and this distinction may likely have some degree of interaction with cancer risk. Generally, 80 % of patients who seek bariatric surgery are female.

Each of these surgical procedures present with specific risks (mortality, perioperative, and long-term complications) and benefits (degree of weight loss and postsurgical lifestyle) and as a result, careful consideration should be given prior to matching the patients' clinical needs and health risks with the potential health benefits as well as surgical and postsurgical health risks. Differences in cancer protection among the different surgical procedures are not known.

The first reported recognition that weight loss through bariatric surgery may have some link with subsequent cancer, in this case mortality, came from a bariatric surgery paper published by MacDonald et al. [49], in which 154 post-bariatric surgery patients who prior to surgery were diagnosed with type 2 diabetes were compared with 78 severely obese type 2 diabetes patients who had not undergone RYGB

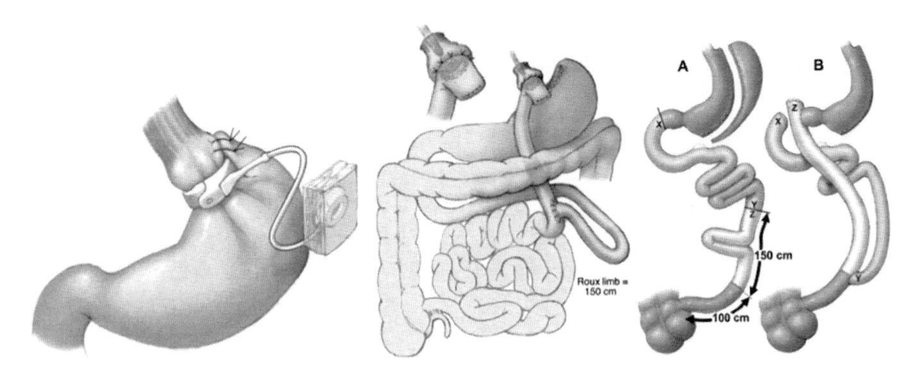

Fig. 9.1 Diagram of bariatric surgical procedures. (**a**) Laparoscopic adjustable gastric band procedure. (**b**) Roux-en-Y gastric bypass. (**c**) (A) Sleeve gastrectomy and (B) Duodenoileostomy (constructed 250 cm from the ileocecal valve with a common channel of 100 cm). When (A) and (B) are performed together, the procedure is referred to as a Duodenal switch (Figures reprinted by permission of © Mayo Foundation for Medical Education and Research. All rights reserved)

surgery [49]. The two groups were reasonably matched with regard to age, sex, and BMI, and follow-up time (RYGB patients followed for a mean of 9 years and controls for a mean of 6.2 years). MacDonald et al. reported a 0 % cancer mortality in the RYGB group compared to a 0.6 % cancer mortality in the controls, which were not significantly different [49]. Since this first report by MacDonald et al., subsequent papers have been published linking bariatric surgery to either cancer incidence or cancer mortality [42, 49–57]. In addition, a study by McCawley et al. reported the relationship of female cancers related to bariatric surgery in women whose cancer had been diagnosed prior to their having bariatric surgery as well as women free of cancer before surgery but diagnosed with cancer following bariatric surgery [58].

Since 2009, there have been five reviews exploring the potential relationship between bariatric surgery and subsequent cancer risk [22, 40, 59–61], and two additional reviews of cancer risk associated with either weight loss from bariatric surgery or nonsurgical weight loss therapies [11, 20]. In addition, Gagné et al. published a retrospective review of prospectively collected patients who were diagnosed with cancer before, during, or after bariatric surgery [62], and Renehan published a brief review comparing key cancer and bariatric surgery studies [63]. From these reviews, this chapter highlights three reported studies related to cancer incidence, cancer mortality, and total mortality following bariatric surgery: Sjöström et al., Christou el al., and Adams et al. [52, 53, 55].

The Swedish Obesity Subjects study (SOS study) is a prospective, controlled study that involved 2010 participants who underwent bariatric surgery (71 % females) matched using several parameters with 2,037 participants who did not undergo weight loss surgery. Study participants were enrolled at 25 surgical departments and 480 primary health care centers in Sweden and of the surgical group, 376 (18.7 %) underwent nonadjustable or adjustable gastric banding, 1,396 (68.1 %) had vertical banded gastroplasty, and 265 (13.2 %) Roux-en-Y gastric bypass procedures [64]. Inclusion criteria included age between 37 and 60 years and a BMI of 34 kg/m^2 or more for men and 38 kg/m^2 for women. The initial SOS mortality study followed participants in both groups for an average of 10.9 years and vital status was determined for all but three of the participants (follow-up rate of 99.9 %) [42]. A significant strength of this mortality study was the prospective tracking of weight. Maximum weight loss from baseline that occurred over the period of up to 15 years was 25 %, 16 %, and 14 %, respectively for gastric bypass, vertical banded gastroplasty, and gastric banding, with an approximate ±2 % weight change among the control group [42]. The unadjusted overall total mortality HR in the surgery group when compared to the control group was 0.76 (95 % CI, 0.59–0.99; $p=0.04$) and when adjusted for sex, age, and risk factors, the HR was similar at 0.71 ($p=0.01$). During this follow-up period, there were 129 deaths (6.3 %) among the control group and 101 deaths (5.0 %) in the surgical group. The most common cause of death over this mean 10-year period was from cancer (48 deaths in the control groups compared to 29 deaths in the surgical groups), with myocardial infarction being the second leading cause of death (25 deaths among the control group and 13 deaths in the surgical group) [42]. Based upon this study, Sjöström et al. published a follow-up paper on the incidence of cancer following bariatric surgery among SOS study participants with a mean follow-up of 10.9 years (range from 0 to 18.1 years) [53]. Reported cancers subsequent to bariatric surgery were 117 cancers in the surgical group compared to 169 cancers among the control groups, representing an HR of 0.67 (95 % CI 0.53–0.85; $p=0.0009$). For female participants only, the surgical group had a reported 79 cancers compared to 130 cancers in the control females, giving an HR value of 0.58 (95 % CI 0.44–0.77; $p=0.0001$). There were no effects related to bariatric surgery and subsequent cancer incidence when the males only were compared (38 cancer cases among men in both the surgical and control groups) [53]. Interestingly, the extent of weight loss or changes in energy intake among the SOS subjects participating in the bariatric surgery group were not significantly related to the reduction in cancer incidence [63]. However, sagittal trunk diameter (a substitute measure for intra-abdominal adiposity [53, 65]) was shown to contribute significantly to cancer incidence [53].

In 2006, Christou et al. published an observational study of weight loss following bariatric surgery of 1,035 patients (65.6 % female; operated on between 1986 and 2002). Surgical patients were compared to a control group ($n=5{,}746$) of age- and gender-matched severely obese patients who were identified from a large health care claims database (which included hospitalizations) using ICD codes related to obesity [50]. The types of bariatric surgery included open Roux-en-Y gastric bypass (7≥ %), vertical banded gastroplasty (18.7 %), and laparoscopic Roux-en-Y gastric bypass (8 %) procedures. Christou et al. reported a mortality rate for the surgical group to be 0.68 % compared to 6.17 % for the control group. The mean follow-up time was approximately 2.5 years (maximum of 5 years) [50]. This study was followed by a second report published by Christou et al. in which first-time physician/hospital visits that linked to an eventual "all cancer diagnosis" were determined for 1,035 post-bariatric surgical patients between 1986 and 2002 [55]. For the purpose of controls, similar to Christou's earlier study, an age- and gender-matched morbidly obese group ($n=5{,}746$) of participants indentified from 1986 to 2002 using ICD codes for morbid obesity, who had not undergone bariatric surgery and whose data were part of a single-payer administrative database was used. If a surgical or control participant was found to have visited a physician or hospital for purposes that were related to cancer (diagnosis or treatment) within 6 months before their inclusion into the study, they were excluded from the analysis [22]. After a maximum of 5-year follow-up, the reported number of visits to the physician/hospital that led to a cancer-related diagnosis for the weight loss surgical group was 21 visits (2.0 %) compared with 487 visits (8.5 %) among the control group. This difference was reported to have a relative risk of 0.22 (95 % CI 0.14–0.35; $p=0.001$) [55]. For the specific cancers related to the physician/office visits, the relative risk for breast cancer was 0.17 (95 % CI 0.01–0.31; $p=0.001$). Whether or not these breast cancers were pre- or postmenopausal was not indicated. For colorectal cancer, the RR was 0.32 (95 % CI 0.08–1.31; $p=0.63$) when the surgical group was compared to the controls. This study did not determine cancer-related mortality among participants of either group [22].

In a retrospective cohort study, Adams et al. determined the long-term mortality (from 1984 to 2002) of 7,925 post-Roux-en-Y gastric bypass patients to 7,925 severely obese control subjects who had applied for a Utah driver's license. Participants in the two groups were matched for age, sex, body mass index, and the date of bariatric surgery was matched to the year the control participant applied for their driver's license [51]. Because of potential driver's license reporting bias with regard to height and weight, the self-reported height and weight of the post-bariatric surgery patients whose driver's license had been renewed less than 5 years prior to their gastric bypass surgery (average 1.8 years; $n=592$; 68 % female) was compared to their clinical records (height and weight before bariatric surgery). Regression equations of self-reported and measured BMI (male and female specific) were then used to correct the self-reported BMI of all driver's license applicants. In addition, all severely obese driver's license applicants selected as controls were compared to the surgical patient's file and to a state registry of all Utah hospitalizations to eliminate the possibility of the control subject having participated in bariatric surgery. Any control subject who was matched to these two sources (i.e., had previously undergone bariatric surgery) was eliminated [51]. All participants of the surgical and nonsurgical groups were submitted to the National Index bureau for the purpose of obtaining mortality status and cause of death. The BMI of the surgical group was significantly less than the control group (45.3 kg/m^2 versus 46.7 kg/m^2, $p<0.001$). Mean follow-up was 7.1 years (18 total years). The number of all-cause deaths in the surgery group was 213 with 321 deaths among the control group, representing a hazard ratio of 0.60 (95 % CI, 0.45–0.67; $p<0.001$) after covariate adjustment. Prevalent cancers for the surgical group were 1.67 % and 1.59 % in the control group ($p=0.71$). For cancer deaths, the bariatric surgical group was 60 % lower when compared the control group ($p=0.001$; 31 deaths among surgical group compared to 73 deaths in control groups). These results equate to 5.5 cancer deaths per 10,000 person years in the surgical groups vs. 13.3 cancer deaths per 10,000 person years among the control group. Analysis of cancer included excluding deaths from cancer that had resulted during the first 5 years following baseline and eliminating prevalent cancers at baseline [51]. A weakness of this study is that weight at the time of death was not available and only self-reported baseline weight was available for the control participants, although these self-reported weights were adjusted by sex-specific regression equations. There was a significant reduction in cancer risk in the bariatric surgical group compared to severely obese controls occurring after a mean follow-up of only 7.1 years. It was acknowledged that reduction in body weight may have the effect of improving the sensitivity of cancer screening, possibly leading to earlier detection of cancer with subsequent improved survival [51]. It was speculated, however, that even potential improvements in cancer screening were insufficient to account for the meaningful reduction in cancer death among the surgical group compared to controls. As a result of the reduction in cancer mortality, cancer incidence was next examined among these two groups of participants.

The follow-up period was extended to a total of 24 years (mean, 12.5 years) [52]. Participant data were linked to the Utah Cancer Registry (UCR) only for gastric bypass patients who were residents of Utah (6,596 of the total 9,949 postgastric bypass patients). A total of 9,442 severely obese control participants identified through Utah driver's license applications were also submitted to the UCR. The driver's license applicant data were group matched to represent the age, sex, and BMI distribution of the surgical patients. In addition, the driver's license self-reported BMI (determined from self-reported height and weight as part of the license application process) was adjusted using sex-specific regression equations [52]. For each participant of the surgery and control groups, the UCR was used to identify the following cancer information: site (type), stage, date of diagnosis, vital status, and date of death according to SEER standards. The SEER staging system codes were defined as follows: 0 = in situ; 1 = localized; 2 = regional, direct extension only; 3 = regional, regional lymph nodes only; 4 = regional, direct extension and regional lymph nodes; 5 = regional, not otherwise specified (NOS); 7 = distant metastasis; and 9 = unstaged [66].

Over the 24-year follow-up period, there were 254 (3.1/1,000 person years) and 477 (4.3/1,000 person years) incident cancers detected in the postgastric bypass and control groups, respectively [52]. Highlighted in Table 9.1 are the incident cancers by site as well as the number of cancers, rates, hazard ratios, and p values. For all cancers combined, there was a 24 % reduction in cancer incidence among the surgical group compared to controls (HR 0.76, 95 % CI, 0.65–0.89; $p = 0.0006$). Whereas all incident cancers were significantly lower for females of the surgical group compared to control females (HR 0.73, CI 0.62–0.87; $p = 0.0004$), there were no significant group differences in incidence of all cancers when males only were compared. However, only 14 % and 17 %, respectively, of the surgical and control participants were men. Interestingly, while incident cancers "likely" to be obesity related were significantly lower in the surgery group compared to controls (HR = 0.62, 95 % CI 0.49–0.78), the nonobesity-related incident cancers were not significantly different between groups. In this analysis, obesity-related cancers were defined as esophageal adenocarcinomas, colorectal, pancreas, postmenopausal breast, corpus and uterus, kidney, non-Hodgkin lymphoma, leukemia, multiple myeloma, liver, and gallbladder. Based upon these analyses, it was estimated that about 71 gastric bypass surgeries would be necessary to prevent one incident cancer [52].

To further explore the reduced incidence of cancer, stratified comparisons by stage of cancer (stage when cancer was first diagnosed) were performed. There were no stage differences between groups in the in situ (stage 0) and local (stage 1),

Table 9.1 Cancer incidence[a] and hazard ratios in the study groups (1984–2002) for common cancer sites, cancers by sex, obesity-related cancers, and nonobesity-related cancers [Table reprinted by permission of the Obesity Society, All rights reserved. Adams et al. Obesity (2009) 17, 796–802]

	Surgery $N=6{,}596$		Control $N=9{,}442$			
Cancer site[b]	Number of cases	Rates/1,000 person years	Number of cases	Rates/1,000 person years	Hazard ratio[c] (96 % CI)	*P* value
All cancers	254	3.13	477	4.28	0.76 (0.65–0.89)	0.0006
All cancers, male	39	3.73	65	3.83	1.02 (0.69–1.52)	0.91
All cancers, female	215	3.04	412	4.36	0.73 (0.62–0.87)	0.0004
Obesity-related cancers[d]	104	1.28	253	2.27	0.62 (0.49–0.78)	<0.0001
Nonobesity-related cancers[e]	150	1.85	224	2.01	0.91 (0.73–1.12)	0.37
Oral cavity and pharynx (20010–20100)	3	0.04	9	0.08	0.46 (0.12–1.75)	0.25
Esophagus (21010)	3	0.04	4	0.04	0.98 (0.21–4.66)	0.98
Stomach (21020)	2	0.02	2	0.02	1.70 (0.24–12.2)	0.59
Small intestine (21030)	1	0.01	4	0.04	0.38 (0.04–3.47)	0.39
Colorectal (21041–51052)	25	0.31	52	0.47	0.70 (0.43–1.15)	0.15
Liver (21071)	1	0.01	1	0.01	1.69 (0.10–27.80)	0.71
Gallbladder (21080)	0	0	2	0.02	–	–
Pancreas (21100)	9	0.11	8	0.07	1.75 (0.66–4.63)	0.26
Other digestive (21130)	0	0	1	0.01	–	–
Larynx (22020)	1	0.01	2	0.02	0.81 (0.07–9.03)	0.87
Lung and bronchus (22030)	5	0.06	11	0.10	0.71 (0.25–2.08)	0.53
Other respiratory (22010, 22020, 22050)	1	0.01	3	0.03	0.51 (0.05–4.97)	0.56
Trachea (22060)	0	0	2	0.02	–	–
Soft tissue including heart (24000)	4	0.05	4	0.04	1.13 (0.26–5.00)	0.87
Melanoma of skin (25010)	17	0.21	29	0.26	0.71 (0.38–1.34)	0.30
Other nonepithelial skin (25020)	4	0.05	2	0.02	2.37 (0.40–14.0)	0.34
Breast (26000)	73	0.90	107	0.96	0.91 (0.67–1.24)	0.54

(continued)

Table 9.1 (continued)

Cancer site[b]	Surgery $N=6,596$ Number of cases	Rates/1,000 person years	Control $N=9,442$ Number of cases	Rates/1,000 person years	Hazard ratio[c] (96 % CI)	*P* value
Premenopausal female breast	49	0.60	65	0.58	0.93 (0.63–1.37)	0.69
Postmenopausal female breast	24	0.30	40	0.36	0.96 (0.57–1.63)	0.89
Cervix uteri (27010)	9	0.11	14	0.13	0.88 (0.37–2.08)	0.78
Corpus and uterus NOS (27020–27030)	14	0.17	98	0.88	0.22 (0.13–0.40)	<0.0001
Ovary (27040)	7	0.09	19	0.17	0.19 (0.23–1.34)	0.19
Vulva (27060)	9	0.11	2	0.02	6.15 (1.30–29.2)	0.02
Prostate (28010)	17	0.21	17	0.15	1.71 (0.87–3.36)	0.12
Urinary bladder (29010)	4	0.05	3	0.03	1.98 (0.43–9.06)	0.38
Kidney and renal pelvis (29020)	11	0.14	13	0.12	1.22 (0.54–2.78)	0.63
Brain and CNS (31010–31040)	6	0.07	10	0.09	0.69 (0.24–2.02)	0.50
Thyroid (32010)	10	0.12	20	0.18	0.64 (0.24–1.41)	0.27
Hodgkin's lymphoma (33011–33012)	1	0.01	2	0.02	0.73 (0.06–8.78)	0.80
Non-Hodgkin's lymphoma (33041–33042)	7	0.09	17	0.15	0.54 (0.22–1.37)	0.20
Myeloma (34000)	2	0.02	4	0.04	0.46 (0.06–3.28)	0.44
Leukemia (35011–35043)	4	0.05	7	0.06	0.37 (0.08–1.64)	0.19
Other	4	0.05	8	0.07	–	–

CI 95 % confidence interval, *CNS* central nervous system, *NOS* not otherwise specified

[a]Incidence is defined as cancers diagnosed after gastric bypass surgery for the surgical group and following application for driver's license or identification card for the severely obese controls

[b]The Surveillance Epidemiology and End Results (SEER) site-specific coding is included within the parentheses. Menopause assumed to occur at age 50 for breast cancer analysis

[c]Analyses are adjusted for sex, age, and BMI. For gender-specific sites, analyses are adjusted for age and BMI

[d]Obesity-related cancers included esophageal adenocarcinomas, colorectal, pancreas, postmenopausal breast, corpus and uterus, kidney, non-Hodgkin lymphoma, leukemia, multiple myeloma, liver, and gallbladder

[e]All cancers that are not included as obesity-related cancers

but regional cancers (stages 2–5) were significantly lower in the surgical group compared to controls (HR 0.61, 95 % CI 0.43–0.89; $p=0.009$). The distant cancers (stage 7) were also significantly lower in the surgical patients compared to control participants (HR 0.61, 95 % CI 0.39–0.96; $p=0.03$). Finally, the cancer case-fatality rates were neither significantly different between groups nor were the mean times to cancer detection [52]. Baseline prevalence also did not differ between groups. Unlike the SOS study, this study only had baseline weight available. Further discussion related to strengths and weaknesses of this study have been previously reviewed [52]. This study further surmised that:

> "...regional and distant cancers that would have resulted without the surgery [gastric bypass] were detected in the in situ and local stages and in situ and local stage cancers that would have occurred without surgery were prevented or delayed beyond the end of the follow-up period."

9.4 Childhood and Adolescent Obesity Prevention and Treatment

A thorough review of the prevention and treatment of childhood and adolescent obesity through lifestyle and medical therapy, bariatric surgery and the potential impact these treatments might have on cancer incidence in adulthood is beyond the scope of this chapter. However, brief highlight of the health burden and eventual cancer development risk of obesity as related to this population is important to note. Over the past 30 years, childhood and adolescent obesity in the USA has increased three- to six-times, with increased incidence dependent upon the age, gender, and ethnicity [67, 68]. Increased disease risks specific to children and adolescents who are obese include asthma, type 2 diabetes, nonalcoholic fatty liver disease, and associated cardiovascular disease-related risks [68–70]. Thompson et al., tracking the National Health and Growth Study (NGHS) population to adulthood, reported children with reported onset obesity prior to age 12 years were 11–30 times more likely to present with obesity as adults. In addition to increased obesity risk, the overweight NGHS children had greater incidence of hypertension, hyperlipidemia, and metabolic syndrome as adults [71, 72].

The potential link between obese children and adolescents and cancer risk would appear to be projected to consequences this population may face as they attain adulthood. Several chapters of this book have thoroughly illustrated the association between obesity and cancer in adults, and so it stands to reason that obese children and adolescents may be at increased lifetime cancer risk due to the evidence indicating that obese children are far more likely to be obese adults. One specific example of this possibility relates to potential future risk for breast cancer among obese female children, where overweight and obesity have been shown to be associated with the earlier age onset of puberty [71]. As a consequence of earlier age menarche, the risk for breast cancer in adulthood is increased [71, 73–75]. Potential biochemical and environmental mechanisms related to obesity and earlier age onset of puberty, including the notion that earlier age puberty may result in an increased "total lifetime estrogen burden." [71, 76]. In a review by Jasik and Lustig, the authors point out that while "it is clear that nutritional status plays an essential role in the timing and progression of puberty ... [it] is not clear whether the weight gain comes before the early puberty, if early puberty predisposes to abnormal weight gain, or both" [71]. In any case, the need to provide children and adolescents with healthy lifestyle exposure, including proper nutritional offerings and adequate opportunity for regular physical activity is paramount in the effort to prevent overweight and obesity and eventual risk of cancer [77]. Further, screening for obesity in children and adolescents with subsequent referral to intensive counseling and behavioral interventions has been recommended by the US Preventive Services Task Force (USPSTF) for the purpose of improving weight status [68]. Similar to previous discussion in this chapter related to possible reduced cancer risk in adults as a result of voluntary weight loss, it would appear that the opportunity for overweight and obese children and adolescents to improve weight status would likewise reduce their lifetime risk of developing cancer.

9.5 Potential Mechanisms Related to Bariatric Surgery and Reduced Cancer Risk

In a recent review article by Ashrafian et al. [40] proposed mechanisms leading to a decrease in cancer tumor generation as well as growth after bariatric surgery were identified [40]. Figure 9.2 by Ashrafian et al. illustrates the concept that bariatric

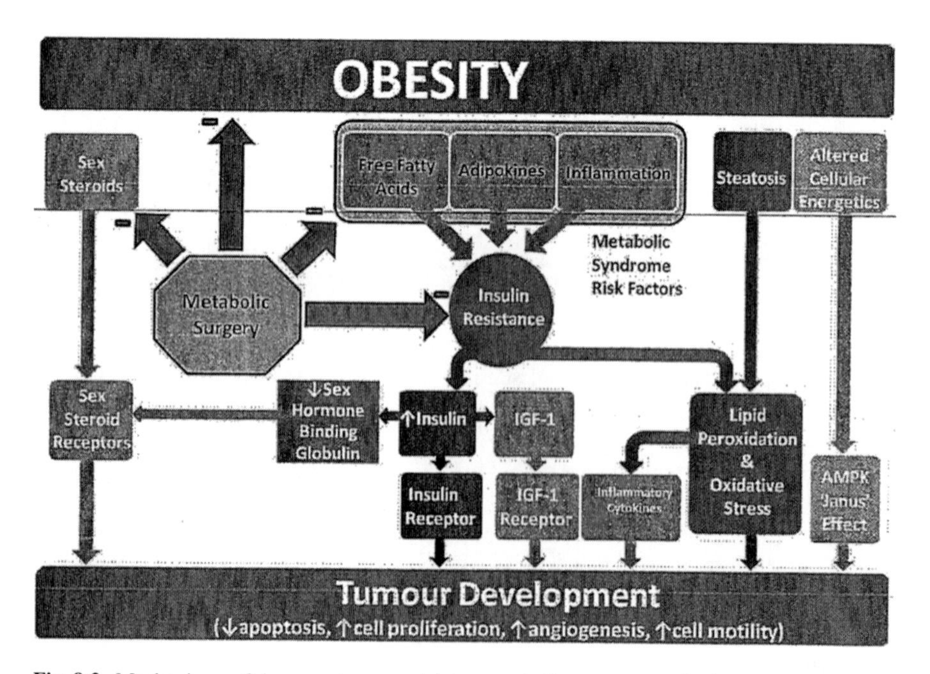

Fig. 9.2 Mechanisms of decreased cancer risk by metabolic surgery are depicted. *IGF-1* Insulin-like Growth Factor 1, *AMPK* 5′ adenosine monophosphate-activated protein kinase [Figure reprinted by permission of John Wiley and Sons. All rights reserved. Ashrafian et al. Cancer. 2011; 17(9):1788–1799]

(or metabolic) surgery "interrupts" the proposed pathways that contribute to both obesity and cancer development.

In very simplistic terms, Fig. 9.2 proposes that due to surgically induced weight loss, there is a reversal in the direction of primary biological mechanisms that link obesity (excess adipose tissue) with increased risk for cancer development such as chronic inflammation, increased sex-steroid hormones, and increased insulin resistance (topics that are carefully reviewed in previous chapters). We emphasize "simplistic" used in the previous sentence. Whereas previously, the physiological reasons for successful weight loss following bariatric surgery was attributed almost exclusively to restriction and/or malabsorption of the gastrointestinal system, current understanding is that in addition to decreased caloric intake there are multiple proposed mechanisms related to "metabolic" surgery [40, 78, 79]. The remission of diabetes following bariatric or metabolic surgery is an excellent example of the multiple and complex mechanisms that have been proposed and studied. From a prospective study examining the effects of gastric of gastric bypass surgery on long-term morbidity, our group has reported a near 80 % remission of diabetes 2 years following gastric bypass surgery [80]. Multiple reviews have been published on the effect of metabolic surgery on diabetes, including a meta-analysis by Buchwald et al., which reported a 78.1 % remission of diabetes and an 86.6 % improvement or remission in diabetes following bariatric surgery [81]. The intriguing element related to diabetes remission is that a significant number of bariatric surgical patients (i.e., gastric bypass patients) have discarded their antidiabetic medication and returned to a normal blood glucose by the time they are discharged from the hospital following their metabolic surgery (i.e., 2–3 days after surgery) and long before significant weight loss has occurred [82]. Again, mechanisms accounting for this remarkable remittance or improvement of diabetes following surgery are multiple. In an analogous way, the reduced risk of cancer following metabolic surgery is also likely to be linked with several biological mechanisms, which may or may not be directly associated with weight loss.

9.6 Items for Consideration Regarding Metabolic Surgery and Subsequent Cancer Risk

Given the relatively few studies published to study the association between metabolic surgery and cancer mortality and prevalence, there remains a meaningful opportunity for additional cancer and bariatric surgery research. Potential questions worthy of investigation include:

- Does the impact of bariatric (metabolic) surgery upon subsequent cancer risk vary in response to the type of bariatric surgical procedure? Bariatric operations differ widely with regard to anatomical and/or mechanistic approaches (i.e., alteration in normal intestinal function) and as a result, neural, biochemical, and biomolecular responses to the surgery can also vary. In addition, the degree of adipose tissue loss and the rapidity of weight loss are different between procedures, which may result in variations of how the procedure affects hyperinsulinemia, a potential link between obesity and the risk for cancer [63].

- Somewhat related to the previous item, what are the mechanistic causes for cancer risk (increased or decreased) following surgery (i.e., weight loss or not, changes in central adiposity, reduced energy intake, menopausal status, etc.)?
- Following metabolic surgery, how rapidly do the effects from surgery influence cancer development (or prevention)? That is, what is the general timeline associated with these mechanisms and does this timeline change as a result of surgery type?
- Does surgery affect specific cancers? If yes, what are the cancers more likely to be impacted as a result of metabolic surgery?
- If a greater number of males were to participate in metabolic surgery, thereby increasing the total "*n*," would this improve the power for detecting change in cancer incidence in males?
- A related question, do protective cancer effects (physiological mechanisms) following bariatric surgery vary based upon gender?
- Does bariatric surgery influence cancer recurrence? If yes, are the mechanisms involved in cancer recurrence risk reduction the same as those implicated in cancer incidence risk? Building upon these questions, how long should a patient be required to wait following active cancer treatment before being cleared for bariatric surgery (i.e., might bariatric surgery have a favorable impact upon postcancer outcomes)?
- What are the potential opportunities to combine robust animal models of obesity and cancer risk with human bariatric surgical outcomes?
- Will future studies that do not rely on self-reported weight data and that include greater prospective clinical data provide greater confirmatory support for metabolic surgery and reduced cancer risk?
- What impact might guidelines related to screening for obesity in children and adolescents with subsequent intensive counseling and behavioral interventions for improving weight status (USPSTF recommendations) have upon lifetime cancer risk?
- Specific to the previous question, how does the magnitude of change in cancer risk resulting from weight loss in childhood and adolescence compare to cancer risk change when weight loss is confined to voluntary weight loss in adulthood?

9.7 Conclusion

Previous chapters have clearly demonstrated the link between obesity (excess adipose tissue) and cancer risk. Less conclusive has been whether or not voluntary weight loss reduced the risk of cancer incidence or cancer recurrence, primarily because of the difficulty in attaining meaningful weight loss amounts and sustaining weight loss in population-based studies. However, limited clinical studies using intense lifestyle intervention have been successful in reducing mean weight for a period of time (i.e., 2–4 years), and a few observational studies have demonstrated reduced cancer incidence following volitional weight loss. Because individuals participating in bariatric or metabolic surgery lose significant amounts of weight and sustain a meaningful degree of weight loss for several years, this population of patients has made possible the opportunity to study long-term weight loss effects upon cancer mortality and cancer incidence. Although also limited in number, there are now studies that demonstrate a reduction in cancer mortality among post-bariatric patients compared to severely obese, nonoperated controls. In addition, one prospective study (SOS study) and a few observational studies have also demonstrated a reduction in cancer incidence following metabolic surgery. To date, the reduced cancer risk benefits have been limited to females and there appears to be a stronger correlation of benefit associated with cancers that are "likely" to be obesity related. Given these limitations, the general consensus is that intentional weight loss does lead to a reduction in cancer incidence [11]. Further, national and international guidelines that recommend individuals lose weight, if clinically indicated, for the purpose of reducing cancer incidence risk appear to be evidence based.

Acknowledgments Kenneth Adams, Ph.D., the late Eugenia Calle, Ph.D., Paul Hopkins, M.D., Richard Gress, M.S., Nan Stroup, Ph.D., Sherman Smith, M.D., Steven Simper, M.D., and Rodrick McKinlay, M.D. T.D. Adams receives partial funding through the *Huntsman Fellowship—Advancing Community Cancer Prevention*, Intermountain Research and Medical Foundation, Intermountain Healthcare Corporation, SLC, UT.

References

1. Chagnon YC, Perusse L, Weisnagel SJ, Rankinen T, Bouchard C. The human obesity gene map: the 1999 update. Obes Res. 2000;8(1):89–117.
2. National Institutes of Health (NHLBI). Clinical guidelines on the identification, evaluation, and treatment of overweight and obesity in adults: the evidence report. Obes Res. 1998;6(2):51S–209.
3. National Institutes of Health. Overweight, obesity, and health risk. National Task Force on the Prevention and Treatment of Obesity. Arch Intern Med. 2000;160(7):898–904.
4. Kushi LH, Doyle C, McCullough M, et al. American Cancer Society Guidelines on nutrition and physical activity for cancer prevention: reducing the risk of cancer with healthy food choices and physical activity. CA Cancer J Clin. 2012;62(1):30–67.
5. American Institute for Cancer Research. AICR. Food, nutrition, physical activity, and the prevention of cancer: a global perspective. Washington DC: AICR: World Cancer Research Fund/American Institute for Cancer Research; 2007.
6. Boyle P, Autier P, Bartelink H, et al. European code against cancer and scientific justification: third version (2003). Ann Oncol. 2003;14(7):973–1005.
7. Lichtenstein AH, Appel LJ, Brands M, et al. Diet and lifestyle recommendations revision 2006: a scientific statement from the American Heart Association Nutrition Committee. Circulation. 2006;114(1):82–96.
8. Bantle JP, Wylie-Rosett J, Albright AL, et al. Nutrition recommendations and interventions for diabetes: a position statement of the American Diabetes Association. Diabetes Care. 2008;31 Suppl 1:S61–78.

9. US Department of Agriculture and the US Department of Health and Human Services. The 2010 dietary guidelines for Americans. Washington, DC: US Department of Agriculture and the US Department of Health and Human Services; 2010.
10. US Department of Health and Human Services. Physical activity guidelines for Americans. Washington, DC: US Department of Health and Human Services; 2008.
11. Birks S, Peeters A, Backholer K, O'Brien P, Brown W. A systematic review of the impact of weight loss on cancer incidence and mortality. Obes Rev. 2012;13(10):868–91.
12. Fisher BL, Schauer P. Medical and surgical options in the treatment of severe obesity. Am J Surg. 2002;184(6B):9S–16.
13. Wolin KY, Colditz GA. Can weight loss prevent cancer? Br J Cancer. 2008;99(7):995–9.
14. Knowler WC, Barrett-Connor E, Fowler SE, et al. Reduction in the incidence of type 2 diabetes with lifestyle intervention or metformin. N Engl J Med. 2002;346(6):393–403.
15. Pi-Sunyer X, Blackburn G, Brancati FL, et al. Reduction in weight and cardiovascular disease risk factors in individuals with type 2 diabetes: one-year results of the look AHEAD trial. Diabetes Care. 2007;30(6):1374–83.
16. Wadden TA, West DS, Neilberg RH, et al. One-year weight losses in the Look AHEAD study: factors associated with success. Obesity. 2009;17(4):713–22.
17. Knowler WC, Fowler SE, Hamman RF, Diabetes Prevention Research Group, et al. 10-year follow-up of diabetes incidence and weight loss on the Diabetes Prevention Program Outcomes Study. Lancet. 2009;374:1677–86.
18. Ryan DH, Espeland MA, Foster GD, et al. Look AHEAD (Action for Health in Diabetes): design and methods for a clinical trial of weight loss for the prevention of cardiovascular disease in type 2 diabetes. Control Clin Trials. 2003;24(5):610–28.
19. Wing RR. Long-term effects of a lifestyle intervention on weight and cardiovascular risk factors in individuals with type 2 diabetes mellitus: four-year results of the Look AHEAD trial. Arch Intern Med. 2010;170(17):1566–75.
20. Byers T, Sedjo RL. Does intentional weight loss reduce cancer risk? Diabetes Obes Metab. 2011;13(12):1063–72.
21. McTiernan A, Irwin M, Vongruenigen V. Weight, physical activity, diet, and prognosis in breast and gynecologic cancers. J Clin Oncol. 2010;28(26):4074–80.
22. Adams TD, Hunt SC. Cancer and obesity: effect of bariatric surgery. World J Surg. 2009;33(10): 2028–33.
23. Calle EE, Kaaks R. Overweight, obesity and cancer: epidemiological evidence and proposed mechanisms. Nat Rev Cancer. 2004;4(8):579–91.
24. Parker E, Folsom A. Intentional weight loss and incidence of obesity-related cancers: the Iowa Women's Health Study. Int J Obes. 2003;27:1447–52.
25. Rodriguez C, Freedland SJ, Deka A, et al. Body mass index, weight change, and risk of prostate cancer in the Cancer Prevention Study II Nutrition Cohort. Cancer Epidemiol Biomarkers Prev. 2007;16(1):63–9.
26. Trentham-Dietz A, Nichols HB, Hampton JM, Newcomb PA. Weight change and risk of endometrial cancer. Int J Epidemiol. 2006;35(1):151–8.
27. Harvie M, Howell A, Vierkant RA, et al. Association of gain and loss of weight before and after menopause with risk of postmenopausal breast cancer in the Iowa women's health study. Cancer Epidemiol Biomarkers Prev. 2005;14(3):656–61.
28. Calle E. Obesity and cancer. In: Hu F, editor. Obesity epidemiology, chapter 10. Oxford: Oxford University Press; 2008. p. 196–215.
29. Radimer KL, Ballard-Barbash R, Miller JS, et al. Weight change and the risk of late-onset breast cancer in the original Framingham cohort. Nutr Cancer. 2004;49(1):7–13.
30. Eng SM, Gammon MD, Terry MB, et al. Body size changes in relation to postmenopausal breast cancer among women on Long Island, New York. Am J Epidemiol. 2005;162(3): 229–37.
31. Webb P. Commentary: weight gain, weight loss, and endometrial cancer. Int J Epidemiol. 2006;35(1):301–2.

32. Williamson DF, Pamuk E, Thun M, Flanders D, Byers T, Heath C. Prospective study of intentional weight loss and mortality in overweight white men aged 40-64 years. Am J Epidemiol. 1999;149(6):491–503.
33. Yaari S, Goldbourt U. Voluntary and involuntary weight loss: associations with long term mortality in 9,228 middle-aged and elderly men. Am J Epidemiol. 1998;148(6):546–55.
34. Rapp K, Klenk J, Ulmer H, et al. Weight change and cancer risk in a cohort of more than 65,000 adults in Austria. Ann Oncol. 2008;19(4):641–8.
35. Ahn J, Schatzkin A, Lacey Jr JV, et al. Adiposity, adult weight change, and postmenopausal breast cancer risk. Arch Intern Med. 2007;167(19):2091–102.
36. Eliassen AH, Colditz GA, Rosner B, Willett WC, Hankinson SE. Adult weight change and risk of postmenopausal breast cancer. JAMA. 2006;296(2):193–201.
37. Decensi A, Puntoni M, Goodwin P, et al. Metformin and cancer risk in diabetic patients: a systematic review and meta-analysis. Cancer Prev Res (Phila). 2010;3(11):1451–61.
38. Freedman DS, Khan LK, Serdula MK, Galuska DA, Dietz WH. Trends and correlates of class 3 obesity in the United States from 1990 through 2000. JAMA. 2002;288(14):1758–61.
39. Sturm R. Increases in clinically severe obesity in the United States, 1986-2000. Arch Intern Med. 2003;163(18):2146–8.
40. Ashrafian H, Ahmed K, Rowland SP, et al. Metabolic surgery and cancer: protective effects of bariatric procedures. Cancer. 2011;117(9):1788–99.
41. Buchwald H, Oien DM. Metabolic/bariatric surgery worldwide 2008. Obes Surg. 2009;19(12): 1605–11.
42. Sjostrom L, Narbro K, Sjostrom CD, et al. Effects of bariatric surgery on mortality in Swedish obese subjects. N Engl J Med. 2007;357(8):741–52.
43. Mingrone G, Panunzi S, De Gaetano A, et al. Bariatric surgery versus conventional medical therapy for type 2 diabetes. N Engl J Med. 2012;366(17):1577–85.
44. Schauer PR, Kashyap SR, Wolski K, et al. Bariatric surgery versus intensive medical therapy in obese patients with diabetes. N Engl J Med. 2012;366(17):1567–76.
45. Dixon JB, O'Brien PE, Playfair J, et al. Adjustable gastric banding and conventional therapy for type 2 diabetes: a randomized controlled trial. JAMA. 2008;299(3):316–23.
46. Buse JB, Caprio S, Cefalu WT, et al. How do we define cure of diabetes? Diabetes Care. 2009;32(11):2133–5.
47. Kushner RF, Noble CA. Long-term outcome of bariatric surgery: an interim analysis. Mayo Clin Proc. 2006;81(10 Suppl):S46–51.
48. National Institute of Health. NIDDK. Bariatric surgery for severe obesity. Bethesda, MD: National Institutes of Health, National Institute of Diabetes and Digestive and Kidney Diseases (NIDDK); 2011.
49. MacDonald Jr KG, Long SD, Swanson MS, et al. The gastric bypass operation reduces the progression and mortality of non-insulin-dependent diabetes mellitus. J Gastrointest Surg. 1997;1(3):213–20.
50. Christou NV, Sampalis JS, Liberman M, et al. Surgery decreases long-term mortality, morbidity, and health care use in morbidly obese patients. Ann Surg. 2004;240(3):416–23.
51. Adams TD, Gress RE, Smith SC, et al. Long-term mortality after gastric bypass surgery. N Engl J Med. 2007;357(8):753–61.
52. Adams TD, Stroup AM, Gress RE, et al. Cancer incidence and mortality after gastric bypass surgery. Obesity (Silver Spring). 2009;17(4):796–802.
53. Sjostrom L, Gummesson A, Sjostrom CD, et al. Effects of bariatric surgery on cancer incidence in obese patients in Sweden (Swedish Obese Subjects Study): a prospective, controlled intervention trial. Lancet Oncol. 2009;10(7):653–62.
54. Peeters A, O'Brien PE, Laurie C, et al. Substantial intentional weight loss and mortality in the severely obese. Ann Surg. 2007;246(6):1028–33.
55. Christou NV, Lieberman M, Sampalis F, Sampalis JS. Bariatric surgery reduces cancer risk in morbidly obese patients. Surg Obes Relat Dis. 2008;4(6):691–5.
56. Ostlund MP, Lu Y, Lagergren J. Risk of obesity-related cancer after obesity surgery in a population-based cohort study. Ann Surg. 2010;252(6):972–6.

57. Gusenoff JA, Koltz PF, O'Malley WJ, Messing S, Chen R, Langstein HN. Breast cancer and bariatric surgery: temporal relationships of diagnosis, treatment, and reconstruction. Plast Reconstr Surg. 2009;124(4):1025–32.
58. McCawley GM, Ferriss JS, Geffel D, Northup CJ, Modesitt SC. Cancer in obese women: potential protective impact of bariatric surgery. J Am Coll Surg. 2009;208(6):1093–8.
59. Basen-Engquist K, Chang M. Obesity and cancer risk: recent review and evidence. Curr Oncol Rep. 2011;13(1):71–6.
60. Menendez P, Padilla D, Villarejo P, Menendez JM, Lora D. Does bariatric surgery decrease gastric cancer risk? Hepatogastroenterology. 2012;59(114):409–12.
61. De Roover A, Detry O, Desaive C, et al. Risk of upper gastrointestinal cancer after bariatric operations. Obes Surg. 2006;16(12):1656–61.
62. Gagne DJ, Papasavas PK, Maalouf M, Urbandt JE, Caushaj PF. Obesity surgery and malignancy: our experience after 1500 cases. Surg Obes Relat Dis. 2009;5(2):160–4.
63. Renehan AG. Bariatric surgery, weight reduction, and cancer prevention. Lancet Oncol. 2009;10(7):640–1.
64. Sjostrom L. Surgical intervention as a strategy for treatment of obesity. Endocrine. 2000;13(2): 213–30.
65. Sjostrom L, Lonn L, Chowdhury B, et al. The sagittal diameter is a valid marker of visceral adipose tissue volume. In: Angel A, Anderson H, Bourchard C, Lau D, Leiter L, Mendelson R, editors. Recent advances in obesity research VII. London: John Libbey & Co. Ltd; 1996. p. 309–19.
66. Johnson CC, Adamo M. The SEER program: coding and staging manual 2007. Bethesda, MD: National cancer institute; 2007.
67. Wang Y, Beydoun MA. The obesity epidemic in the United States–gender, age, socioeconomic, racial/ethnic, and geographic characteristics: a systematic review and meta-regression analysis. Epidemiol Rev. 2007;29:6–28.
68. Barton M. Screening for obesity in children and adolescents: US Preventive Services Task Force recommendation statement. Pediatrics. 2010;125(2):361–7.
69. Mokdad AH, Serdula MK, Dietz WH, Bowman BA, Marks JS, Koplan JP. The spread of the obesity epidemic in the United States, 1991-1998. JAMA. 1999;282(16):1519–22.
70. Reilly JJ, Methven E, McDowell ZC, et al. Health consequences of obesity. Arch Dis Child. 2003;88(9):748–52.
71. Jasik CB, Lustig RH. Adolescent obesity and puberty: the "perfect storm". Ann N Y Acad Sci. 2008;1135:265–79.
72. Thompson DR, Obarzanek E, Franko DL, et al. Childhood overweight and cardiovascular disease risk factors: the National Heart, Lung, and Blood Institute Growth and Health Study. J Pediatr. 2007;150(1):18–25.
73. Petridou E, Syrigou E, Toupadaki N, Zavitsanos X, Willett W, Trichopoulos D. Determinants of age at menarche as early life predictors of breast cancer risk. Int J Cancer. 1996;68(2): 193–8.
74. Titus-Ernstoff L, Longnecker MP, Newcomb PA, et al. Menstrual factors in relation to breast cancer risk. Cancer Epidemiol Biomarkers Prev. 1998;7(9):783–9.
75. Stoll BA. Western diet, early puberty, and breast cancer risk. Breast Cancer Res Treat. 1998; 49(3):187–93.
76. Lustig RABH. Obesity and cancer. In: Yeung SEC, Gagel RF, editors. Internal medicine care of cancer patients. New York, NY: Dekker; 2008.
77. Uauy R, Solomons N. Diet, nutrition, and the life-course approach to cancer prevention. J Nutr. 2005;135(12 Suppl):2934S–45.
78. Ashrafian H, Bueter M, Ahmed K, et al. Metabolic surgery: an evolution through bariatric animal models. Obes Rev. 2010;11(12):907–20.
79. Ashrafian H, le Roux CW. Metabolic surgery and gut hormones—a review of bariatric entero-humoral modulation. Physiol Behav. 2009;97(5):620–31.
80. Adams TD, Pendleton RC, Strong MB, et al. Health outcomes of gastric bypass patients compared to nonsurgical, nonintervened severely obese. Obesity (Silver Spring). 2010;18(1):121–30.

81. Buchwald H, Estok R, Fahrbach K, et al. Weight and type 2 diabetes after bariatric surgery: systematic review and meta-analysis. Am J Med. 2009;122(3):248–256 e245.
82. Schauer PR, Burguera B, Ikramuddin S, et al. Effect of laparoscopic Roux-en Y gastric bypass on type 2 diabetes mellitus. Ann Surg. 2003;238(4):467–84. discussion 484–465.
83. Ligibel J. Obesity and breast cancer. Oncology (Williston Park). 2011;25(11):994–1000.
84. Protani M, Coory M, Martin JH. Effect of obesity on survival of women with breast cancer: systematic review and meta-analysis. Breast Cancer Res Treat. 2010;123(3):627–35.
85. Patlak M, Nass SJ. The role of obesity in cancer survival and recurrance: workshop summary. Washington, DC: The National Academies Press; 2012.
86. Hewitt M, Greenfield S, Stovall E. From cancer patient to cancer survivor: lost in the transition. Washington, DC: Institute of Medicine and National Research Council; 2005.
87. Demark-Wahnefried W, Platz EA, Ligibel JA, et al. The role of obesity in cancer survival and recurrence. Cancer Epidemiol Biomarkers Prev. 2012;21(8):1244–59.

Index

M.G. Kolonin (ed.), *Adipose Tissue and Cancer*, DOI 10.1007/978-1-4614-7660-3,

Printed by Books on Demand, Germany